Time, Memory, and Society

Recent Titles in
Contributions in Sociology

Homelessness in the United States: State Surveys (Volume I)
Jamshid A. Momeni, editor

Explorations in the Understanding of Landscape: A Cultural Geography
William Norton

Postmodern Social Analysis and Criticism
John W. Murphy

Suburbia Re-examined
Barbara M. Kelly, editor

Impossible Organizations: Self-Management and Organizational Reproduction
Yohanan Stryjan

Religious Politics in Global and Comparative Perspective
William H. Swatos, Jr., editor

Invisible Victims: White Males and the Crisis of Affirmative Action
Frederick R. Lynch

Theories of Ethnicity: A Critical Appraisal
Richard H. Thompson

Population and Community in Rural America
Lorraine Garkovich

Divided We Stand: Class Structure in Israel from 1948 to the 1980s
Amir Ben-Porat

A Fragile Movement: The Struggle for Neighborhood Stabilization
Juliet Saltman

The Sociology of Agriculture
Frederick H. Buttel, Olaf F. Larson, and Gilbert W. Gillespie, Jr.

The Urban Housing Crisis: Economic, Social, and Legal Issues and Proposals
Arlene Zarembka

TIME, *Memory*, and **Society**

FRANCO FERRAROTTI

Contributions in Sociology, Number 91

GREENWOOD PRESS
New York • Westport, Connecticut • London

Library of Congress Cataloging-in-Publication Data

Ferrarotti, Franco.
[Ricordo e la temporalità. English]
Time, memory, and society / Franco Ferrarotti.
p. cm.—(Contributions in sociology, ISSN 0084–9278 ; no. 91)
Translation of: Il ricordo e la temporalità.
Includes bibliographical references.
ISBN 0–313–26828–2 (lib. bdg. : alk. paper)
1. Time. I. Title. II. Series.
BD638.F4313 1990
115—dc20 89–27370

British Library Cataloguing in Publication Data is available.

Library of Congress Catalog Card Number: 89–27370
ISBN: 0–313–26828–2
ISSN: 0084–9278

First published in 1990

Greenwood Press, Inc.
88 Post Road West, Westport, Connecticut 06881

Printed in the United States of America

The paper used in this book complies with the Permanent Paper Standard issued by the National Information Standards Organization (Z39.48–1984).

10 9 8 7 6 5 4 3 2 1

TIME, *Memory*, and **Society**

FRANCO FERRAROTTI

Contributions in Sociology, Number 91

GREENWOOD PRESS
New York • Westport, Connecticut • London

Library of Congress Cataloging-in-Publication Data

Ferrarotti, Franco.
[Ricordo e la temporalità. English]
Time, memory, and society / Franco Ferrarotti.
p. cm.—(Contributions in sociology, ISSN 0084–9278 ; no. 91)
Translation of: Il ricordo e la temporalità.
Includes bibliographical references.
ISBN 0–313–26828–2 (lib. bdg. : alk. paper)
1. Time. I. Title. II. Series.
BD638.F4313 1990
115—dc20 89–27370

British Library Cataloguing in Publication Data is available.

Library of Congress Catalog Card Number: 89–27370
ISBN: 0–313–26828–2
ISSN: 0084–9278

First published in 1990

Greenwood Press, Inc.
88 Post Road West, Westport, Connecticut 06881

Printed in the United States of America

The paper used in this book complies with the Permanent Paper Standard issued by the National Information Standards Organization (Z39.48–1984).

10 9 8 7 6 5 4 3 2 1

Contents

Preface

Time's reasons cannot be easily accommodated within the framework of traditional historicism. Leonard Krieger is certainly correct in trying to defend the autonomy of history against the widespread claim that history is coming to an end. His book, *Time's Reasons*, is both a passionate attempt and a well-reasoned argument aimed at restoring the autonomy of history as an academic discipline. The question remains, however, and looms larger than ever before: which history? History as an élite enterprise or history "from below," that is, history of the everyday? Conceived as a self-enclosed intellectual effort, history has given birth to historicism, of which there are many uses, as it has been often pointed out (see, for example, the recent book by David D. Roberts, *Benedetto Croce and the Uses of Historicism*). But the tension between historical time, individual and collective memory, and social context in its multifaceted transformations seems to be a permanent characteristic, not to mention the extreme variety of its linguistic expression. Moreover, through remembrance, historical time becomes contemporary time, and is constantly reenacted and relived through a process of "remembering interiorisation," which is essential to the constitution of the subject. Using Dilthey and Bergson as personalized polemical targets, this book intends to show the constructive importance of past experience in order to discover and recover the meaning of present day living and of the seminal values of tradition.

Time, Memory, and Society

The Map of Time

- Time
  - a) Biographical Existential (Unilinear)
    - Life History
    - Life Story
    - Autobiography
  - b) Historical (Unilinear)
    - Macrohistorical
      - Long Duration
      - Generation
      - Event
    - Politicoinstitutional
      - Tradition
      - Revolution
      - Evolution
    - Economicoconjunctural
      - Business Cycles
      - Relations of Production
        - Individual Production
        - Production Line
        - Piece Work
      - Labor Organization
  - c) Ahistorical or Sacred (Cyclical)
    - "Festa"
    - Mythic
    - Hierophanic
  - d) Natural (Cyclical)
    - Daytime
    - Nighttime
    - Seasonal

Time, Memory, and Society

The Map of Time

- Time
  - a) Biographical Existential (Unilinear)
    - Life History
    - Life Story
    - Autobiography
  - b) Historical (Unilinear)
    - Macrohistorical
      - Long Duration
      - Generation
      - Event
    - Politicoinstitutional
      - Tradition
      - Revolution
      - Evolution
    - Economicoconjunctural
      - Business Cycles
      - Relations of Production
        - Individual Production
        - Production Line
        - Piece Work
      - Labor Organization
  - c) Ahistorical or Sacred (Cyclical)
    - "Festa"
    - Mythic
    - Hierophanic
  - d) Natural (Cyclical)
    - Daytime
    - Nighttime
    - Seasonal

Introduction

One says: the passage of time; time flies; *ruit hora.* It is as if I were watching the flux of time, standing still on a riverbank, at a distance. I am watching time flow by. Common sense and conventional wisdom support my illusion. In reality, I am watching myself. It is I who am passing by. Time passes in me, in me and with me. As the great Borges profoundly intuited, "To deny the succession of time, deny the Ego, deny the astronomical universe, these are hopeless appearances, and secret consola tions. . . . Time is the material of which I am made. Time is a river which drags me along, but I am the river; it is a fire which consumes me—but I am the fire. The world, unfortunately, is real; I, unfortunately, am Borges."[1]

Why concern oneself with time, a peculiarly refractory object, and until recently relatively alien to rigorous sociological analysis—a reality one does not disturb, which one does not see, and which does not seem available for smelling, hearing, or tasting?

This is far from a novel argument. It well expresses the attitude of cautious, disenchanted, if not downright ironical skepticism which led the doctors of nineteenth-century paleopositivism to search feverishly for the soul by cutting up corpses. But precisely for this reason, the choice of time as subject for reflection is not arbitrary. It follows the general plan of my work, after *Storia e storie di vita* (History and Life

Histories),[2] but beginning with *La piccola citta* (*The Small Town*).[3] Even earlier, sociological analysis seemed to me a human undertaking concerning human situations as given and experienced. As a consequence, no researcher can declare him self off-limits or imagine himself as external to the research process without paralyzing it. All this time, temporality appeared to me as the essential fluid dimension of the social. As such it is necessarily connected with the biographical and autobiographical approach. It is a required itinerary in that it is the only way to reintroduce the dynamic moment of social reality as a mental construct into the analytical process reified by the technique of poll sampling and generally by the inevitable limits of quantitative methods. This involves, along with the moment of social reality, sense perception and symbolic mediation as the flow of evolution realized in its temporality, and the whole of its meanings gradually expressed.

The biographical and autobiographical approach, rooted in time, allows the eliciting of problematic human situations and distinguishes radically between the technical problem—resolvable through the exact practical application of a given, ideally indifferent, and so interchangeable formula—and the human problem. The latter is specific, not reducible, unrepeatable, seemingly not solvable, external, and essentially mechanical. Rather, it remains and reasserts itself as a problem, an open, permanent tension. This means that the usual way of laying out and proceeding with sociological analysis must be decisively inverted. First, qualitative methods bring out and facilitate the original expression of the problem. Only at a second stage will quantitative methods intervene to verify or falsify as an under-laborer, so as to check some aspects quantitatively, aspects appearing as in themselves quantifiable. The order of priorities in the research process normally followed in the human sciences tends, because of its rigid structure and the opaqueness of the techniques it uses, to encourage the loss of the sense of the problem through the crude postulation of technical and human problems as equivalents.

The critical questioning behind this research moves in three directions:

- against the freezing of the living and its embalming through the application of quantitative methods; against the shortcomings of paleopositivism, quanto-phrenia, the illusion that the facts have an intrinsic theoretical autonomy and thus "speak for themselves"; against the use of rigid questionnaires, coldly constructed and me chanically administered, cutting out the specific (historical) characteristics of the context and those of the specific human situation being examined
- against essentialism, or the absorption of specific human experience, within the timeless frameworks of abstract conceptualization, which presumes to consume in its own aseptic, claimed universalism, in itself, human experience with its specificity dated and historically determined
- in favor of the conception and practice of social research as a human encounter, an exchange of meaningful information, a co-research; or, as acceptance of the other subject, and reciprocity the results of which are not in fact nor a priori open to being taken for granted

Finally, it should not be necessary to warn that my reflection takes shape as a moment which, by analyzing time, involves the sleepwalking of the everyday, that is, unreflecting temporality, that of the animal, the customary, whereby life is lived as a drowsiness, according to a sequence of mechanical gestures and behaviors in the framework of a general theoretical purpose which fuses the everyday to history.

NOTES

1. J. L. Borges, "Nuova confutazione del Tempo," in *Altre inquisizioni* (Milan: Feltrinelli, 1963), p. 186.
2. Published in Rome–Bari by Laterza, 1981.
3. Published in Milan by Comunita, 1959.

1

The Remembrance

MARCUS AURELIUS'S TRAVELER

This book is a sequel to other books, especially to *Il paradosso del sacro* (*The Paradox of the Sacred*)[1] and *La storia e il quotidiano* (*The History and the Everyday*). It is a reflection aloud, uninterrupted, though not without digressions, parentheses, and changes of course. It is also a surrender to an inner impulse, a stage in the difficult apprenticeship of living with a problem. "I pity youth, as I pity what mine was, happy as it was. That little brother of 17, who was me. Poor little brother. When I look back, scared as much by my imprudence then as by my cowardice. How could I have dared! And how could I not have dared! The eternal flow and ebb of the sea on the beach, for a whole life-time."[2]

There is a moment in life, for me, when to live no longer seems enough. One needs to look back and to survey the present situation coolly. In this instant, I suspend the adventure. I attempt to strike a balance. I blacken pages. I am not living. I am writing. I am not living. I foresee living, or relive the past. Meanwhile, I am not living. I float between memory and plan. I relive yesterday's experience and project myself into an anticipation of life. In any case, I am waiting to live. I mortgage the present with a view to the future. Tomorrow is at a premium on today. I postpone life. This is the emigrant's complex: when

I have made my fortune, then I'll go back and start living. Now I confine myself, forced by the mortgage, to work.

Is there any value in concerning oneself with the meaning of one's life or the meaning of history? Why not calmly accept the sleepwalking of the everyday and its constraints as customary, as ancient wounds no longer to be noticed? Can one all at once withdraw from remembrance and project, almost playing one against the other? Just thinking outside and beyond the usual coordinates of predatory, demonstrative-exhibitionist thought. Letting oneself think in an undeliberated, unplanned, unwilled attitude. Not trying to demonstrate why everything exists and has no need of demonstration in order to exist.

Marcus Aurelius's traveler forgets the purpose of his journey along the way. But one can also imagine a traveler who does not follow a precisely drawn route, who follows no path. One can imagine a traveler who walks across meadows and streams, obeying the impulses of the moment rather than following the directions of a well-drawn map; a *flâneur*, not a salesman. But Marcus Aurelius's traveler cannot escape temporality. He can avoid a predetermined itinerary, but he cannot escape time without petrifying himself into an absolute attitude, without thus giving himself over to inertia. Moreover, he can have no self-awareness except by recalling and reliving past experience, and thus without recapturing it as a whole in the global spatiality which would be his internal terrain, the space, supposedly inviolate, of his intimacy.

This, then, is the paradoxical situation where my daily adventure is being played out: a future which has no future and a past which never passes. The future has already always begun; it is the present. The past is what we are because we are what we have been. More exactly, *we are what we remember we were.*

In May, my father would watch the sky with apprehension in case the rain should come after the grass was cut. In a period of a few days the mown grass might be reduced to bedding straw, the meadows to dung heaps. Over the long winter, the livestock might starve. The same apprehensive gaze was directed to the torrid sky of late June and July. The hot sun of

St. Louis's day (Aug. 25) could not win over the deep-rooted distrust of my father. Concern for May's rain would give way to fear of devastating summer hail. In between went the waving golden wheat. Along with feed for the animals, the family's bread was at stake. Time as climate and time as succession of days and weeks, months, years, meteorological conditions, and temporal sequence seemed united; they were within a unique existential experience. Meteorology and chronology were not yet separated. With natural logic they expressed the basic unity of the living—its organic growth, development, aging, and death.

In Piedmont in May and again in June there was the mowing of the meadows. The occasion could never be predicted with certainty. It depended on the earliness or lateness of the season, measured by my father's stern stare. His yellow eye reflected the landscape like a mirror aged in the pallor of the forgotten weak lights lit in the kitchen during the war. From time to time, up at five, this man would give orders to the day-laborers, spy on the quiet swarming of the peasants, give furtive, questioning glances toward the strip of sky lightening to the east from the byre and the stable where he went in and out feeding the animals.

In the farmer's eye there was control, authority, accepted dependency; but there was also music. There was rhythm in that look, the order of movement: the gesture at the right moment, things in their place, Novalis's "arcane night" which makes a cosmos germinate and grow from chaos. Father, in May and June, watched the sky apprehensively. In any case, the unforeseen, the unmanageable was the weather: "Will it rain today? What will it do tomorrow?" And all in one, "Shall we have time?" In the ancestral peasant experience, time as climate and time as succession of days and nights, meteorological and chronological, helped each other, converged, and were entwined.

Meteo-time and chrono-time were unified by husbandry. Once the grass in the meadows was mown and the yellow wave scythed, if it did not dry and were not then at once removed and set aside, if by misfortune it was surprised by a summer downpour, in a few days it was changed into a dung heap. From hay to compost. Principles of decomposition and fermentation. The stinking sewage gave off marsh gas and flickering will o' the wisps in the long summer night. Life began again from the muddy water-meadows in silence. Happily the compost (*laetamen*) fed the bare fields later on, in the smell of the autumn smoke. Smoking black mounds under a sky full of pitch-black crows made a chessboard of the white winter countryside waiting for the life-cycle to start again.

To prevent the hay from becoming compost there took place the passage from atmospheric to chronological time. The peasant began to take on the feverish rhythm of the piece-worker. He discovered haste, the race against time. His sadness, in the evening, revealed a betrayed ideal: to live in unison with both atmospheric and chronological time, to overcome the contradiction between the two. As a child, I sat motionless in the dark corridor watching the universe dancing for hours in the ray of sun filtering through cracks in the door.

Weather time, chronological time, and magical time, sacred in the true sense; time out of time, epochal, the suspension of the everyday. More exactly, a contemplation of the everyday so serene and absorbed as to calmly bring out its essential meaning in all its indeterminate interconnections with the movements of life, with the totality of the meaningful moments of a life not wholly and only resolving itself on the quantitative level of a homogeneous, smooth series of experiences enfolded between a birth and a death. We cannot escape from time, but common language deceives us and leads us into the error of reifying time as a "thing," an object we have or do not have, spend, or save, like a prison which confines us. We say "I have time"; "I have no time now"; "When I have time." However, time is not external to us. It flows within us. It is the rhythm of our heartbeat, the music in our veins. We cannot escape it for the good reason that *we ourselves are time*. The discords, contrasts, and disjunctures between the times must be explored because as a whole they are the evolutionary law of life, the logic of development of biorhythms and sociorhythms.[3] They help us achieve the difficult transition from life story to sense of life.

The norms linked to the colors of a traffic light regulate

automobile and human mobility. But the situation is more complicated than it seems at first sight. In Rome one can move on the yellow as well as on the red. In Naples red is only a vague indication of danger. However, when the green light finally shows, Roman drivers, perhaps Neapolitan ones still more, hesitate; they wait, keeping their foot on the brake. They are paralyzed, as though taken aback or off-balance. They are making sure no one on the other side passes on red. It is a question of rhythms, of lightning understandings in a situation of generalized guilt complex and bluff. Here one does not drive; one plays poker.

In New York being late for an appointment means losing it, becoming a no-show, unreliable, putting at stake, and at a blow ruining one's professional seriousness and social respectability. But in Greece even today, and in Latin America, being punctual at a meeting is considered offensive; it means a basic lack of faith in one's neighbor.

Societies and the cultures which express their relatively permanent values have their sense of time, as we shall see. Some societies, however, lack this sense almost entirely, as people with no ear for music. But what appears to be a lack of a sense of time is perhaps only a peculiar sense of it. Intercultural comparative research does not abound in this field, despite its importance. One must refer to personal experience, closing an eye to its inevitable idiosyncrasies. Whoever has traveled and stayed in India, even recently and especially in certain areas less touched by Western European-style industrialization, will have experienced a feeling which makes the technologized Western man giddy. This is the sense of a mysterious timelessness, a kind of acceptance of that which still seems in unison with the harmony of the cosmos, a being "out of time" (or past-bound) while, inevitably, living in time. We thus find "passist" societies, closed and immobile in absorbed contemplation of their past, and "futurist" societies so imprudent as not only to open themselves to scientific inquiry, but to plan, with meticulous exactness, the tempo of life which does not yet belong to them. They mortgage a future on which they can as yet stake no right to property.

It is hard and perhaps impossible to distinguish, as has at

times been tried, between quantitative-mechanical-astronomic-physical time and qualitative-experiential-intimate-internalized time. I have already noted how the exaggerated distinction between meteo-time and chrono-time in the real experience of the rural world collapses. No doubt there is an essential difference between using the dawn's light as a gauge for marking the beginning of one's daily work, and consulting one's wristwatch to determine when to go to the factory or office. But it is harder to grasp the qualitative difference between the signal transmitted by the ringing from the village clocktower and that of the factory siren; technically different, the two signals are similar by virtue of their social function. Reflection on the fleeting, elusive nature of time is useful to me. It is a lesson in modesty. It helps me to get away from Cartesian rationalist thought. What is too clear is no longer clear at all. In the end it becomes impoverished. There is something Saturnine in science. Scientific analysis tends to impoverish, if not to strangle and devour its object.

Quantitative methods in social research explain too much; they explain everything. Their tests destroy every margin, even just to imagine alternative or to some degree different explanations. I think I began to doubt current sociology when I realized that sociological explanation functioned too well, a perfectly docile, lubricated mechanism. Tried truths left no margins of indeterminacy. They clipped the wings of possible consciousness. They allowed of no alternative interpretations. They reduced the area of unforeseeability—what there is in the human animal that is human, as against the nonhuman animal. Even the success of current sociology seems basically suspect. The success is what has occurred, the already achieved, the acquired, that which blocks other possible happenings. It predisposes to repetition, and thus to conformism.

I have been saved—if I will save myself—through the taste for wandering, Odysseus's vice. I see in mechanically applied quantitative methods, high-precision instruments which have, however, the inconvenient feature of freezing and naturalizing real human experience, reducing or even annulling its problematic character—not exactly foreseeable and not precisely measurable. They block the emergence of the problem. They are like a skull, where the bones can be stripped and polished, but they do not live. Their false vitality is a half-truth. They pretend to base themselves on critique, but in fact they lack critical spirit and cannot distinguish the essential from the contingent, the foot from the shoe. They have the value of an exercise that is gymnastic rather than formative. One does not think with them, but is restricted to making more or less refined combinations. Their "intelligence" covers their instrumental nature. It helps to forget the commodification of values, of which they are effective and persuasive vehicles. It is the slippery fluorescence of the putrid. Only a rationality, open toward arationality and the not-yet-rationalizable, but already there as a wider rational awareness, a dawning rationality not yet fixed in place, can take into account the evolving complexity which makes up the very essence of the living. There is a pragmatic datum and a problematic one.[4] At their point of intersection the sciences of man and of nature can still recognize and meet each other, so that, perhaps, one day there will be a single science, propelled by the creative asynchronies of multiple times which are, in their specificity, unpredictable.

From this perspective it may seem clear that the modern social sciences must be aware of the need to give themselves a systematic order. They do not tolerate crisis. They seem unable to see their problem through. Crisis lacerates, opens deep rifts in formal logical structures, disorders established schemas, and forces established interpretations to be reviewed. But crisis also forces one to see afresh. It has an epiphanic function. Canonic historicism, old and new functionalism, organicism and systems theory—the schools of sociology are seeking a new armor. Historicism, in its double version of idealism and Marxist-materialism, is seeking a systematic definition at the same time as history is diminishing. Historiography seeks new paths. Marxism seems at once too optimistically voluntaristic and lacking a specific theory of personality. It is at once voluntaristic and impersonal. It makes appointments with historical development which are not kept; it awaits the fall of capitalism which does not arrive. It postulates a "new man" of whom nothing is known. It has penetrated the current mode of thought, has become a philosophy of journalism and of common sense. It is everywhere and not to be found. Attempts to make it "scientific," ahistorical, and atemporal have simply mummified it.

Functionalism has met no better fate, rapidly transformed into a sort of convenient *passe-partout* which explains and justifies everything while justifying nothing. Functions and values end up corresponding to each other with such rigorous precision as to be confused with each other. If the old positivistic mechanicism lacked a sense of becoming and intentionality, neofunctionalism lacks a sense of crisis, struggle, and failure. Tragedy is punctually resolved into comedy. Material interests of life are a distant, rarified region. Individuals do not exist. The "law" floats above everything, as sovereign as it is indefinable.

Sociological explanation is impoverished. The temptation to ideology finds its most radical challenge in Popperian falsification, in the framework of a nondialectical absolute naturalism where demand for the systematic turned into its opposite, in the precise prevalence of the skeptical demand which scented in every systematic construct an ideological trap, dogmatic closure, and subjective preference passed off as intersubjective truth. The attempt to overcome historicism and its poverty without clarifying its elite nature has ended by following ever more mechanical paths with structuralism, organic ones with functionalism, and on the other hand ever more subjective ones with phenomenology and ethnomethodology.

Here it becomes hard to deny the interest in neosystemic efforts to withdraw from the order-disorder dilemma, system or antisystem, and which seem to be aware that nature is not only law or function but also possibility, fluidity, development, decomposition, and reconstruction. Thus one departs from nineteenth-century rationality, from what I have called (with a formula possibly more bombastic than penetrating) "rationalistic rationality." The vision of rationality becomes more complex. It is no longer a question of counterposing, romantically, the reasons of science to those of life. It seems clear today that this is a counterposing without meaning, impoverishing both science and life. From the notion of prescriptive and necessitating law one moves to a notion of law understood as a problematic need for congruity, interconnection, possibility, and further widening in a framework of temporal evolution in which no one seems to have the right to set up unpassable pillars of Hercules.

What I am forced to observe, as against the easy expectations of those who conceive of the development of the sciences according to a rising diachronic line where the best necessarily coincides with the different, is that the fall of the "system," after contributing to the opening of a parenthesis of "polimorphic" sociologies (if not wild ones) has soon shown itself not to be the premise for recovery and reconstruction of a critical sociology. This polimorphic interlude, making one experience briefly the drunkenness of immediate, adventurous encounters with the social, beyond and at times against the blinkered schemas of tradition, did not produce critical sociology open to the demands of society and sensitive to history as broad, many-sided historical life. From the "system" we passed rapidly to the polymorphic. But thence we descended and, seemingly irredeemably, were entangled in the gratuitousness, unforeseeable and apparently capricious, of the amorphous. This is marked by myriads of little fragmentary studies, not led by penetrating conceptual apparatuses, rhapsodic about their subjects and uncertain in method. Their miscellaneous *ex tempore* nature all too openly betrays their immediacy and the slight seriousness of their sponsors, be they private industries, municipal or regional authorities, unions, or pressure groups. The risk today lies in forgetting or slyly keeping quiet about the fact that the critique of traditionalized theoretical categories necessarily involves a renewed strategy of categorization. There are no palliatives or short-cuts.[5]

Were a new Croce to arrive on today's intellectual scene, he probably would not have much difficulty in doing what he did in *The History of Italy from 1871 to 1915*, even with recent methodological tools and without the rigor of an iron rake needed to clean the Augean stables of a nonoriented sociological study, often illegible and always subordinate to the interests of the moment. The passage from "system," reified and pseudo-unifying, to uneven fluidity was and remains necessary. But in no way can it justify the amorphous confusion of the current gratuitousness. The contributions of philosophy for sociology in today's condition are important not so much as a contribution to the demarcation between the scientific field and metaphysical speculation as, in my view, in modern conditions. These are not those of the Vienna Circle at the beginning of the century. They might show themselves decisive in helping sociology rediscover its internal critical nature, which it needs, and without which it flattens out into ingenious description.

THE PHILOSOPHICAL QUESTION AND SOCIAL RESEARCH

Philosophy can decisively clarify a pair of preliminary moral choices, that is, real metatheoretical presuppositions, which sociological research cannot dispense with. First, there is the need to go beyond more or less strongly commodified predatory thought. This does not mean failing to recognize the relative social determinacy of mental or ideal constructs, but rather taking them out of the logic of the market which tends to replace the criterion of efficiency and short-term returns for that of thinking as a calm exploration and gradual approximation to the description and interpretation of a human situation.

The second metatheoretical presupposition is the broadening of the historical perspective, which involves considering human groups which have no history, and until the present have existed in history but as passive raw material, inert fuel, awaiting the flame from on high to use them for its own ends.

Regarding the essential problems of sociological research on a world level, philosophical reflection is perhaps able to make sociologists understand the basic function of historical consciousness relating to their studies and the very formation of their concepts. Naturally, one should not confuse historical consciousness with historical erudition. But contemporary philosophy would not be critically capable of the task if it accepted idealistically its own dilution and reduction into its own history. No doubt, in fact, every philosophical problem has its history. Equally true, however, is the fact that it is not consumed in this. Even if Comte was far from lacking in historical knowledge, it is today's sociologists above all—aside from some of the classics—who are lacking in historical awareness.

Dialogue with philosophy should be reopened on this subject with extreme openmindedness. What is the meaning of "historical consciousness"? What does historicity in philosophy mean? How is it possible that a sociologist, well-endowed and historically erudite, a classic, should so lack historical awareness as to crudely equate modern democracy, for example, with its "electoral rites," with the cults of Athena and Jupiter in classical antiquity?[6] The selection of facts helps the formation of historical consciousness beyond and above historical information. But on the basis of what criteria of evaluation? Where do values originate, historically changeable but yet like submarines emerging seemingly unharmed in different, even conflicting and violently antithetical, historical epochs?

Clarifying the concept of historicity, philosophy gives considerable aid to sociology. First, the relative nature of any sociological categorization appears. The very "laws" of social development, when one does not have the heart to renounce this late nineteenth-century terminology, seem acceptable only in the sense of tendential uniformities or broadly probabilistic predictions. Not only this, but historically equipped sociological research, aided by philosophy in obtaining an adequate historical consciousness, even when it enters into the field of intercultural comparative research, should not give credit to hasty cross-cultural homogenization, superimposed synthetic globalizations, or dogmatic unifications, which necessarily relate back to an impossible ultimate foundation.

To restrict myself to a recent example, a study like Almond and Verba's on civic culture and political participation in five different countries in North America and Europe would not be possible and at all events would not be long in showing the mechanical way in which the concept of political participation has been adopted and used, without taking into account the different historical backgrounds which give the same linguistic term different meanings and, it would seem, meanings incompatible without the right weightings. From this there derives a further service of philosophy for sociological research; the preliminary declaration of value premises from which one starts and the taking on of an explicit viewpoint corresponding to an act of self-placement both historical and ideal, on the researcher's part, and a renouncing at the same time of atemporality and the myth of absolute objectivity.

Thus the way to arrive at a concept of truth as a dialogic conquest is opened up, one thereby intersubjective and contextualized, as well as of the research itself as socializing or coresearch moment—over and above naturalistic or mystifying reifications. At the same time one could recover the lost sense of the problem—a loss which in my view is the most serious lack in contemporary sociology. I have tried elsewhere to reconstruct the itinerary of this lack, certainly linked to an inadequate awareness of many students in the social science field, deservingly desiring to throw themselves into the sea of objectivity and collect the greatest number of empirical data, but unfortunately paying for their acritical formulation of assigned research objects, by assignors who are satisfied with the quantitative precision of research techniques but unconcerned about the interpretative instance, convinced as they are that the facts speak for themselves. The blind alley of a nonproblematic accumulation of data and information with the greatest technical investigation—not knowing for what end—thus becomes inevitable. Equally lamentable and inevitable on the politico-practical level is the offer of social knowledge to the wielders of power and established interests with no chance of control over the use of this knowledge, so as to make sociological research fall into the condition of a mere tool of social spying.

The problematic philosophical demand has largely managed to make sociological research proceed beyond the paleopositivistic stance. "Theory" is today almost universally recognized as an essential moment for orienting research; and integrating theory and research is, at least in principle, respected by researchers as basic to their procedure. Along with the logical-linguistic definition of the terms used in the formulation of the problem to be investigated—often following Merton's guidelines on the paradigm—this is today regarded by sociologists as common ground. But conceptual confusion, with dangerous consequences of a practical order, is still considerable in this regard. It especially concerns the two concepts of theory and model. Theory is mostly reduced to the model and thereby is uprooted from its context and dehistoricized. Doubtless the model is thus more manageable. It is a mental construct, a whole of connected elements in an internal logical congruity, and thus unitary and convergent, with a certain degree of plausibility, even if models remain essentially arbitrary, ahistorical "units of reason." Theory always refers back to questions of substance. It is an outlook (*theoreìn*), a specific outlook, and thus one on a determinate reality. As such, theory can never be removed from the tension between the demands of a given society or the needs of life and the material and ideal relations of human beings in a determinate context, and the cognitive apparatuses and operational modes corresponding to them.

The confusion between the concepts of theory and model, especially the reduction of theory to model, gives rise to a process of dehistoricizing sociological analysis which paradoxically heralds its triumph and at the same time its essential failure. The "system" and structural-functionalism then appear once more, after the polymorphic phase and the amorphic interlude, in an extreme version which sanctions the end of history in the name of the self-referential system. This promises the reduction of the complexity and governability of social demands through their systematic amputation, the cancelling of every possible consciousness and thus any extrasystemic potential. Finally, they reject as laughable any social function of utopia. No doubt the end of history opens up to sociological research other paths which on cursory examination may appear very promising: from sociobiology to symbolic interactionism, from dehistorized Marxism in a neoscientistic key to a semiology which reduces the social to an amusing interactionist metaphor, from verbal competence as an essentially psychologistic dialogic construct to the contraction of all human experience into an aesthetic-epiphanic moment which in essence denies time and history.

The question of what history this involves remains. The diachronic development on which the philosophies of history rested, perhaps more theological than teleological, or Eurocentric history "from above" which saw itself as the last confine of human civilization, is certainly at the end of the incipient synchronic age. Eurocentric history prevented understanding others, did not accept the otherness of others, laying out as the sole possible form of contact forced assimilation or genocide. But history as historical life and pluralistic coexistence of human cultures is scarcely at the beginning, perilously suspended between dawn and sunrise. This is the most precious warning that philosophy today can send to sociological research, too often bent over the immediate datum to take note of new historical horizons, so intent on the fragment that it loses sight of globality. It will be necessary to reexamine the use of operational concepts in sociological research.[7] Further, it will be necessary to go beyond the dichotomic-schematic attitude, where community too conveniently is counterposed to society, structure to history and culture, movement to institution, individual to group and class.

The argument is complex and difficult and offers no hope of peaceful destinations. To concern oneself seriously with the relations between philosophy and sociology can in the end help one understand that things are not so simple as they seem, that sociology is not what is practiced today. The term *sociography* and in some cases *investigative journalism* or *simple impressionistic journalism* are more suitable. Philosophy itself is not, as believed especially in Italy, the history of philosophy but its exact opposite: the ability to question itself and the other disciplines with ingeniousness and obstinacy, almost as though philosophy as a specialized field of knowledge had not yet begun and, indeed, had never existed.

A binary scheme, based on counterpositions of concepts which are mutually exclusive, does not seem adequate to give an account of the variety, if not the richness, of the current scientific debate. It seems necessary to resist the temptation of sharply separating the two contexts in which the cognitive drive of human beings is manifested—that of validation, verification, or falsification, and that of discovery. These are two contexts which, together, make up science. But they are radically different and cannot be reduced one to the other without seriously mutilating scientific activity.

One will have to admit a plurality of stances. In other words, as has been penetratingly remarked, the problem is that of the relation between the systematic rationality of science and the work of research which allows one to arrive at scientific knowledge and make it go forward. Ludovico Geymonat in most of his works complains that today it has become fashionable to speak ill of rigor, that, even among mathematicians, the deeper nature of reality is ignored, a nature varying by way of nuances, changes of viewpoint, inversions of perspective, and the like. It is true that traditionally and crudely, intuition has been contrasted with rigor, vision with demonstration.[8]

It has simply been forgotten or left unsaid that science is both these things. Seen as systematic rationality, science is a rigorous method, a process based on precision of terms and connections between propositions. As such, science makes an abstraction from the historical conditions favoring its birth and conditioning its flowering. On the other hand, seen in the framework of the context of discovery, as exploratory research activity, science is just one of many human activities. It does not proceed only on the basis of the criterion of internal self-correctability, hand in hand with the rise of new data requiring new theoretical formulations, but experiences the normal conditionings and typical setbacks of all human activities. It is not only the history of successive unilinear and harmonious cognitive processes. It is also the history of what Schumpeter called the "original visions" and then the "schools" struggling among themselves, of plots, political clashes, power, and blood.

The two aspects of scientific activity—the systematic rational and the intuitive and social—are both to be borne in mind to avoid the risk of restricting the view of research into the narrow frames of a rationalism unable to appreciate the fertility of crisis and the heuristic potential present in contradictions. No doubt the danger of a fall into irrationalism is real, and the proclaimed superiority of intuition can hide the rejection of analysis and rigorous demonstration. However, today experience appears richer than thought and the Cartesian—and paleopositivist—prejudice seems incapable of making us understand any longer a reality which at all levels has become complex, not to be cut up into facts self-sufficient and closed in themselves, but fluid, magmatic, without design, wandering aimlessly in a fluid, boiling mass.

The category of temporality thus comes to occupy a central position. The mobile perspectives of a permanent evolution, which proceed by doublings and complications and differentiations, enter into fixed logical and epistemological structures. The names of Bergson and Ortega y Gasset return, the vital urge and manifold perspectivism. With them returns the need not to charge the complexity of life with the inadequacy of scientific methods one makes use of to study, understand, evaluate, and foresee it. This, in many ways unexpected return has already found its interpreters both in the field of philosophy in the strict sense and in the field of science. From Michel Serres to Ilya Prigogine human and natural sciences seem to move in an atmosphere of nonbelligerence, if not yet of stable alliance. There is an attempt to develop Bergsonism by applying to space the notions of complexity already worked out for time. The current method of procedure in science and the new style of research make these graftings possible, fruitful, and even stimulating. From being a romantic critic of science, generally put in the company of the most laggardly neoidealists like Croce, for whom the sciences *en bloc* are condemned and downgraded into "inferior means of intellectual life," Bergson is promoted as an anticipator of the current scientific perspective intended to grasp the authentic essence of the movement of reality, which never has the fixity of crystal but if anything is made from fluidities and the uncertain, plastic, and changing nuances of a cloud.

I am aware of the danger of yielding to the cloudy fascination of the confused and the indistinct, the occasional, in the name of a weak thought or one weakened because the thinker does not have the capacity to think things through. But it is true that today the essentially problematic nature of time, of myself in time, of me as time, forces me to give up comfortable rationalistic rationality and suggestive systematics, all-enfolding but empty, a marble sarcophagus for nothing. Their basic support—dogmatically conceived science erected into scientism—diminishes. One understands with the whole man. Reflection and memory mutually nurture each other. In March farmers cleaned the meadows' borders. It was the time of the first cutting of the grass. Weeding expressed the deep, metamechanical sense of the maintaining of the countryside. It was a loving act as well as a necessary pruning, demonstrating attitudes and care which related back to the Georgics of Virgil. Grass was gathered into little mounds and it and the cut branches were burned. Sprinkling around the ash, with straw, new life, invoked recognition of the mortal, finite character of all human undertaking. The gestures were sober, to the rhythm of the days, hours, moons, and seasons. At the end of March, one cleaned up the borders of the land with the meticulous attention of a Genevan watchmaker. The spring rains would not consequently transform the ditches into muddy, threatening torrents. The showers would not become ruinous floods. The anguish of the deadline was not present, the deadline which dominates the rhythms of societies given over to the technical process.

Here there was the apprehensive fatherly stare which presides over birth. Always father, from April to July, apprehensively watching the sky. What will it do tomorrow? Life time and atmospheric time seemed closely united in a strange death pact. The paternal stare is grim because it is threatened by the uncertainty of the event and the thought of Saturn who greedily eats his own sons—but the paternal stare is also threatening as regards the gods whose envy is notoriously a harbinger of death.

The grass waves like a green sea from the ditch, the favorite place for clandestine swims despite the worried remonstrances of anxious mothers. The joyous rediscovery of the amniotic waters of one's origins. The father too was still immersed in the sacred, not because he went to church or took the formal sacraments. Rather, because he cast mournful glances at the sky, and if the clouds were gathering and threatened rain or hail, he murmured between his teeth, with a scarcely audible whisper "*Porcu diu.*" Then with head lowered, closed on himself like a fist, he demanded that God justify himself. No spell, no explanation in terms of the causal nexus could suffice. His anger expressed the direct link with the sacred, with the open sky as church and witness.

What I as a trembling child saw in that paternal look was the metamorphosis of meteo-time into chrono-time, the tragic transition from nature as mother to Saturn, devourer of sons, but also the fluid, ambiguous outline of other times—cyclical and unilinear time, the time of family rites and labor, the time of birth and death, the time of the long fabulous summer and of school days, the time of *festa* and market.

THE PEASANT WORLD AND THE TECHNICAL ENVIRONMENT

According to Georges Friedmann, between the peasant and artisan world on the one hand and industrial society and the technical environment on the other there is a gap which cannot be measured strictly in terms of per capita income and GNP, technical-productive progress per man hour, global production, or even by reference to family groups or the bureaucratic, depersonalized structure of public administration. Between the rural world and the technically advanced environment the essential difference concerns the motion of time. If one understands rhythm as the order of movement, one must recognize that in the preindustrial era "natural rhythms," as Friedmann calls them, prevail. In Western premachine civilization, where the individual depended on the elements, seasons, livestock, the cycle of day/night, and so on, work was constantly subject to the influence of natural rhythms. But it would be a mistake to think that this rhythm is only biologically conditioned. On the contrary, the rhythm of the human body, and especially those of skilled work, have been slowly built up within groups and civilizations; they express secular traditions, at times millenary ones, and are derived from causes not just biological but also sociological.[9]

Friedmann's approach here recalls an extraordinary convergence with the position of the economic and social historian John U. Nef, who, however, seems less interested in the outlining of the contrast between rural world and technical environment, in establishing the origins of the myth of quantitative precision or exactness, so important in a society based on mechanical progress and whose welfare depends on industrialization on a vast scale.

Elsewhere I have taken up Nef's intuitions which may be

connected to Pierre-Maxime Schul's studies and especially Koyré's, as well as Lucien Febvre's *Le problème de l'incroyance au xvi siècle.* In my *The Myth of Inevitable Progress* I tried to establish the meaning of this cult of precision of quantitative exactness in the framework of the logic of mass production, that is, on the basis of the logic of "scientific" organization of labor insofar as this is contrasted with traditional artisan "rule of thumb."

It is in fact true that time as a precise measure is not very important in peasant and artisan civilization. Friedmann has often stressed how the sense of time, or duration, was very different in the rural-artisan world from that in the contemporary industrial one. In Friedmann's judgment, the lack of precise reference points to measure the passing of time and the indifference to this absence—working from dawn until dark, the use of noon or the village bells, slowness (not to be confused with neglect or idleness)—place at the fore the deep pulsing of natural rhythms, from personal biorhythms to the changing of the seasons. Too-demanding timetables, or the unit of measurement as a socially important unit, seem amenable to being left aside. In some Mediterranean countries still today, to be punctual at an appointment can be mistaken for a lack of faith.

Time as a mechanical, quantitative unit is, on the other hand, a highly important datum for the technical environment and urban-industrial society. It is the study of time which makes the scientific organization of work possible, the predetermination of the processes necessary to arrive at the production of any industrial product. Lewis Mumford was so impressed by the importance of time in technically advanced industrial societies that he came to state, overturning traditionally established explanations, that "the watch, not the steam engine, is the key machine of the modern industrial age."[12] But it is again necessary to observe that this importance is not restricted to the world of production: it also affects the concept of duration and what has value. In the peasant world—dominated by the awareness of scarcity, repetitive and static, where the diffuse sense of the precariousness and ultimate impotence of man in the face of nature—what has value is that which lasts, that which endures through time. From this also derives the basic sense of the value of tradition, which cannot be measured nor understood in purely logical terms. Tradition refers to a precious source of psychological assurance and ethical justification of individuals and social groups, but is also at the same time a collection of cognitive and operational data or a technique of survival. In the urban-industrial environment, quality and duration are radically dissociated. In industrially mature societies and—improperly labelled in my view—postindustrial societies, it is actually necessary to produce consumption. Quality is replaced by interchangeability. Using evocative paraliterary testimonies, Friedmann works out a radical counterposition between two worlds, that of nature and that of technique. He presents this dichotomy as valid on a global scale and in all cultures. One should not thereby think that he falls a victim of stereotyped historical schematism. Although aware of simplistic explanations, however, at some points he yields to the temptation of a general picture and evolutionary logic.

> Man is not always the same; he feels, acts and thinks differently in the various epochs of his history, and according to the environment he lives in, according to the techniques he has. Prehistory, now without any doubt, points to the first attempts at tools, the first pieces of stone, of worked flint, . . . clubs, hammers, saws, . . . man modifies his environment and through the environment himself, and moves off towards new transformations. Nothing is continuous in this march, nothing linear, univocal. Civilizations are born and die, some stagnate isolated from the techniques discovered by other human groups: they proceed to their destiny without knowing them, passing as it were into the margins of history. Today this is impossible in the world: technical civilization because of the prodigious means of diffusion it has at its disposal is, in this sense, totalitarian.[13]

It is above all in *Où va le travail humain* that Friedmann maps out the two basic concepts of nature and technique as two realities in contradiction. From these two contradictory realities there descend the two related concepts, symmetrical and opposed, of natural-existential time and mechanico-technical time. Some doubts are justified in this regard. Friedmann bases his theory on the very evocative thought of Georges Navel's *Travaux*, from which Friedmann takes inspiration and examples. Friedmann regrets not having been able to consider Navel's *Parcours*.[14] He is well aware that Navel's works have an essentially psychological tone and content in that they reflect individual experience. In this they recall Willy Helpbach's *L'uomo della metropoli*, as well as the classical remarks by Simmel on mental conditions in metropolitan life. Friedmann recognizes that between the two worlds there are areas of reciprocal interference, if not superimposition, but he does not draw the logical conclusions: "In the various regions of the globe there coexist, in an infinitely varied array, two worlds, two environments of influences and reciprocal relations. In many countries man finds himself caught between them, tossed about, and at times even torn apart."[15]

What concerns Friedmann has also challenged scholars of industrial relations and unionists up to recent times—that is, until the large-scale introduction of computers and robots. Perhaps because of the rather romantic counterposition of the two worlds, of the direct contact between nature and technique, the treatment of human problems of industrial mechanizations Friedmann provides, although certainly stimulating, fails to go beyond a vision of the problem evocative of Chaplin's *Modern Times*. Rather, one must recognize that industrial labor in fragments has been in some way reunified thanks to the internal development of the technical processes of production, especially with the introduction of transfer machines (those controlled by automatic servomechanisms and robotization controlled by computers). Friedmann's attention is almost wholly absorbed by work on the production line, undoubtedly an important phenomenon, but not decisive in advanced industrial development. Friedmann never wholly abandons the idea of a primordial nostalgia which invariably accompanies the ex-peasant worker. Again Navel's *Travaux* is presented as something more than a purely individual experience. In his eyes it takes on the value of an emblematic one. Friedmann writes:

> This experience becomes individual in a curiously gifted personality, but one which is not at all exceptional. How many others, having neither the gift nor the taste nor the strength to express themselves as Navel did, have equally but in silence experienced the drama of

still being nourished and brought up on all the juices of the natural environment, thrown into the hubbub of the great cities and mass production! The joy, the need or the nostalgia for the natural environment form the weave where our worker's experiences continually intermingle, condemned to earn his living in a world for which he is constitutionally unsuited. No sooner has he suffered in Lyons the first disappointments with the ill-ventilated dark workshops, now deprived of the great lyrical hopes of his elders, than he finds refuge and consolation in the countryside, abandoning himself, his hat covered in patches, to the joy of the warm scented air.[16]

In reality, it is true that the notion of the time of the rural and the industrial worlds changes radically, and that work by time organized according to the principles of the scientific organization of labor has little in common with artisan work over the whole process. But the notion of time crosses over the world of work, and should not only be considered on the level of production, in the workshop or the artisan's shop, but rather in the global society. After Friedmann one would say that with industrialization time was rationalized and to some extent simplified. But in reality things are more complicated. And only an ingenious rationalist could think that the great transition from the natural world to the technical environment means a simplification of temporal processes.

On the contrary, one could argue with some justification that the question of time, once not even perceived as a problem, so much did it seem blended with the natural pace of the seasons, emerged with the advent of industrialized societies as a problem in the human, that is, the full, sense—not resolvable with the mere correct application of a given technical formula. It has already been remarked that time was above all a means for orienting oneself in the social world to regulate coexistence among men.

The natural sequences standardized and manipulated by men found application as instruments to determine the position of duration of social activities in the flow of events, and only recently has there been derived from that the use of watches as useful instruments for studying purely natural sequences. The mystery of time, always noticed by man, is thus reinforced.... That watches are instruments produced and used by men for determinate ends and on the basis of certain requirements of their common life is not hard to recognize. But that time too should have an instrumental character, as it appears, is something more tricky to understand. Do we not feel how it passes inexorably over men's heads? Linguistic habit also, moreover, obstructs our judgement. It makes it seem to us that time is something mysterious which, however, watches, instruments created by men, can measure.[17]

The question of time does not thus seem reducible on a first, cursory examination, to the dichotomy between rural world and technical environment, between as it were existential and mechanical time.[18] The problem is complex. This complexity is already suggested, if not explicitly expressed, by language. The linguistic use of the term *time* does not allow any simplistic resolution of the problem. One must note, however, that in some languages, like French and Italian, the same term *temps, tempo* is used to indicate two groups of radically distinct phenomena, that is, atmospheric conditions, what we have called meteo-time, and chronological succession, or chrono-time.

MEMORY AND THE MOMENT

Time is proteoform. It appears to me from the start to be in essence multiple and polymorphic, presenting itself as a secret maze which challenges memory and logical reconstruction and in which men wander about, terrorized and lost or unaware. They are at any rate unable to escape the fatal trap of a repetition which is at once continuous and ever-different, multiple and basically unitary, lightning and everlasting.

I have been. I shall be. What I have been I am no longer. What I shall be I am not yet. I am only what I am here and now, at this point in time, this precise moment. The moment I think I can grasp it and hug it to me, it has already run, passed, ended, died. The experience of the moment is at once dangerous and constitutive for the subject. The I who thinks lives in the moment he thinks. Time is thought in the instant where the thinking I reflects on it. At the same instant, time is already no longer present and not yet future. The time of the instant is both different from the time of duration (in the past) and successive time (in the future).

> The experience of the instant, meeting point and point of conflict between history, maze, repetition, mirror, reached through the dream or return to the same place, is extremely dangerous. We feel lost here, the world participates more than ever in our being and we are a thing, alone with the real. We no longer distinguish between internal and external. We reach such an anomalous state of fusion between thought and being, thought and reality, that Borges speaks explicitly regarding this experience on the one hand of idealism and on the other of death. In one sense, in fact, the absolute realization of idealism, idealism according to Berkeley and Hume for whom no other reality exists save the ideas of the mind, leads to the death of man, the death of that part which sets him apart from other living things. A most strange thought this, explicable only by passing through the experience and analysis of circular time, and thence the experience and analysis of the moment.[19]

The problem is how to exit from the labyrinth of the instant. For we are what we are and know we are that only in the moment of reflection. We are what we are reflected in the glass; the mirror confirms that we are by reflecting us. Every morning I identify myself, shaving, in front of the bathroom mirror. Robert Lowell confirms this: "One morning in the month of July 1954 I was sitting in my bedroom on the third floor of the Payne-Whitney clinic in New York hospital, trying as usual to get my picture of myself straight."[20]

The operation of self-definition may coincide with that of involuntary self-execution. One can always be struck down by the short circuit of the lightning instant. For we are what we have been—more exactly, what we *remember* being. But how is memory possible? The lived is at once mediated by the already experienced.

Every meaningful experience is always an already experienced one, literally relived. Memory is Ariadne's thread which guarantees getting out of the labyrinth and safety. Memory is the ability to remember, to preserve one's traces, interpret fragments, to reconstruct yesterday's world on the basis of tell-tale signs, the bases of identity through the renewed continuity of experience, to rediscover the secret of duration. But memory is a faculty that forgets. Man is a remembering animal, but memory is selective. What to remember? To forget? And why?

A first schematic response is already possible, a provisional template. Memory recalls the crucial moments on which the person is built, the experiences deeply experienced by the person, more exactly, by the personality of the person. For we are all human beings, marked by a common humanity, but at the same time each of us is an individual, an *unicum*, a prototype. First it was not. Then it was. Is. And then again, is no more. When it dies something unique dies, an irreplaceable being, not to be substituted, nor copiable, nor interchangeable. The death of any human being is thus an irreparable loss. The death leaves an unfillable gap. But precisely for this reason we need eternity. Without eternity there is no history; there can be neither biographies nor autobiographies. Without eternity historical history and personal destinies are nonsense. Cicero intimates this when he remarks in the Tusculan meditations that man is the only animal who plants a tree and will not enjoy its fruits.

So, experiences. But which ones? Those which present themselves spontaneously, without being explicitly remembered or recalled? Experiences which confront consciousness involuntarily or those which are consciously treasured? Memory is selective, but the logic of this selectivity is not easily mastered. What makes an experience a deep experience? Not just a lived experience, but one participated in and experienced so radically that it amounts to a transformation?

Memory is selective. It is not the antique shop of a possessive dealer who accumulates objects in confusion and with voracious bulimy consigns everything to its dusty destiny. Memory filters, selects, archives. The absolute memory of Borges's character Juan Luis Ireneo Funes must be an absolute torment:

With a glance we perceive three glasses on a table. Funes, all the tendrils, clusters and fruit of a whole bower. He knew the forms of the austral clouds at dawn on April 30 1882, and could compare them in his memory to the veining on a leather-bound book he had seen only once, and the outline of the spray sent up by an oar on the River Negro on the eve of the Quebracho affair![21]

This memory by way of images—eidetic, so pitiless, precise, and flatly photographic as to prevent any evaluation or reordering of memories according to a scale of priorities, at least relatively—ends up cancelling itself out. But it has the undoubted merit of underlining that the selection of memory, recall, does not respond to a voluntaristic mechanism.

From the neurobiological viewpoint, the selectivity of memory has been shown, or perhaps only intuited, by the research of Gerald Edelman and others. Here Darwinism, with its central principle of selection, has scored a victory as precious as it is unexpected. Against subjectivist or desubjectifying—but mystically introspective (from Dilthey to Bergson and Heidegger)—philosophies, it recalls the decisive moment of the impersonality of the memory together with the discontinuous nature of tradition and the archaic, quite other than harmonious and smooth. Walter Benjamin's cryptic phrase is perhaps here identified in its deepest meaning: "One must define the dialectical image as the *involuntary memory* of redeemed humanity" (emphasis added).[22]

Memory is never only and completely voluntary; it does not merely fall in the sphere of the subjective. It necessarily recalls and refers to the context. Images, fragments of experience, sights and smells of the surroundings, childhood afternoons, the perfume of a first love affair, the terror of school exams emerge unexpectedly, confront consciousness, even involuntarily, determined at least in part or recalled by involuntary memory based on the network of fluid associations of recall or remembrance. Here lies the basic importance of certain pieces of music, certain social and religious functions and monuments, lasting from infancy at the bottom of a consciousness still unconscious of itself, ready to reemerge at the propitious moment which is not necessarily willed by the subject.

In this sense memory is not simply an individual question. It has a base, a link with the community. It is at bottom a community experience. It involves the group, the collective unconscious, a stream of consciousness which links everything

and travels in the interior of everyone at variable speeds and with its own images without, thus, exclusively belonging to anyone. The involuntariness of memory can at times create the impression of an internal chronological disorder (*presque inconscient: qu'il vous échappe*). Images overlay each other, yield, intermingle, and pile on one another in the radiograph of consciousness.[23] Walter Benjamin spoke of a "photographic plate" of consciousness and further "conjured up" (possibly the right word in this case) the method to restore order into the chaos of images, by way of involuntary remembrance. He theorized a kind of historical index of images. This index tells us that the images manage to be read together only in a specific period: "It is exactly this attainment of legibility which is a determinate critical point within them. Every present is determined through those images which are synchronous with it: every *Jetzt* is the *Jetzt* of determinate knowabilities." This means that the past preserves itself as though impressed on a photographic plate, which, however, shows its image only at a specific time which depends on the interrelations with other images which have been revealed. "The past has images deposited in it which can be compared to those fixed on a sensitive plate. Only the future can develop them. Those which are strong enough—so that the image can appear in all its details." But not even the transmission of the past is certain; there is no past which can call itiself "once and for all safe in the granaries of the present."[24]

If the memory has a communitarian existential and experiential basis, thus a historical one in the full sense, memory is not simply an intellectualistic faculty, a mechanism of pure reason—to use the striking formula of Bachelard, this "stranger without memory and without heredity." Memory, on the contrary, arises from a complex matrix. It involves the experience of one's own body, and thus of space, temporality, and the world concretely lived and experienced in their specificity. From being to becoming. The unfolding of time involves the body and its processes of evolution and ultimately of decomposition. Time elects the body as its privileged vehicle, shows itself in its jointings, its heartbeats, its trembling, and its wrinkles. The complexity of memory and its operations essentially derive from the fact that in reality memory is two things, corporeity and consciousness. Subjectivity is temporality which manifests itself in it. And temporality is history in its fullness as history of the community—archaism, tradition, and project.

In the strict sense we imply the social through a synthetic internalization which de- and restructures it by giving it psychological forms at the same time. But as it is produced by a synthetic praxis, the relation which binds an act to a social structure is not linear; the close relation between a social history and a life is certainly not a mechanical determinism. The time dimension is thus not given by the succession of mental states believed to be interconnected by causal nexuses external to consciousness. Rather, consciousness and temporality coincide. Temporality is consciousness in action. It is its field of presence, the modality of its synthesis, the generalized condition of its self-transcendence and self-direction.

So, memory saves, as consciousness and corporeity together, that fluctuating precariousness which we call the subject by long tradition, through the narcissistic temporality of the pleasure principle. At the same time it reinforces awareness and constructs the ego or the personality as reflex sentiment of the continuity of the past (an intrinsically problematic continuity) through the flow of memories which reemerge in the present as representation of the past. But they are also full of anticipation and of auroral promises of the future. Past, present, and future appear at this stage as convenient schematic subdivisions: the past never passes, the present is already future, which in turn already flows back into the past. There is neither past, present, nor future, but only a vast, apparently disordered (because not intellectualistically mastered evocative co-presence of multiple) interlaced times, recall, echoes, presentiments of a sovereign reality where consciousness and memory, nostalgia and project, corporeity and psychic states face each other, mingle, and are based. Selectivity of memory gives us the possibility of reflecting on ourselves and reproducing what we have already been (but still are), so as to proceed to the comparison between past and present in the light of the future. In this sense it denies and preserves us from the Medusa of the magic instant. Given the co-presence of times, it makes us understand how the deepest reality that seems to be offered by this atemporal instant is not only ungraspable but essentially illusory. Thus too the dichotomy between lived time and atemporality collapses. The gods of the moment, who can be grasped only ecstatically, are a gift, not a proposal by man who acts in, lives in, and lives time. Willed or self-imposed atemporality is equivalent to self-annihilation.

The problem is thus not that of living in time, but rather in living time, that is, entering into the dialectical short circuit of experience, memory, tradition, established community values, reappropriation of the collective, self-identification as abstraction, insertion, and individuation. I have already said that every individual is but a singular universe. We as individuals are time; we live it according to diverging and converging perspectives with a plurality of styles which passes over us into a historical horizon where, on the other hand, our presence is essential.

HISTORY AND LIFE IN NIETZSCHE

Nietzsche's "antiquarian history," like Bergson's *durée* and Dilthey's *Erlebnis*, has severe limitations from this point of view.

In *Considerazioni inattuali* (*Untimely Reflections*), especially on the basis of the deep influence exerted on him by Schopenhauer, Nietzsche attacks the prevailing culture of his time, first through the "cultured philistine." He then critically attacks the advocates of historicism who, by reducing every human experience to the dimensions of official history, end up limiting or simply closing the horizons of this experience, impoverishing it both cognitively and emotionally. Hence the title of these early writings *Untimely Reflections*. Untimely because they turn against the culture of timeliness, that is, the current intellectual fashions in the name of a deeper need, linked to the double necessity of understanding the reasons for modern man's decadence and the terms for the reconstruction of man as a meta-human being (Nietzsche's superman). In this sense he speaks of a new morality "beyond good and evil," and of an inversion or overturning of all traditional values. In fact he sees in traditional morality, especially Judaeo-Christian morality, a force which has historically hindered the betterment of the human race through the protection of the weaker and more regressive.

In line with this basic presumption he defines Christian morality elements as a slave morality. One must replace this morality by one of free men, or a morality of strength, will to power, vital energy expanding in full freedom with no limits except those which are self-imposed.

In the second *Reflection* Nietzsche criticizes one of the most important myths of his time (and still partly of ours): the cult of history as expressed and perpetuated in the different forms and schools of historicism. However, one must note that Nietzsche's polemic is never univocal, crude, or dogmatic. On the contrary, Nietzsche's polemic is instrumentalized by political forces and currents of thoughts on opposite sides, from Nazis to idealists, romantics, decadents, existentialists, and Marxists. In the first part, Nietzsche seems both favorable and hostile to history. Does he then contradict himself? In fact he starts by observing that the qualitative difference between man and animal, both natural beings, lies in the fact that the animal does for its own sake, and so repeats itself mechanically in its behavior on the basis of immediate instinctive drives, whereas man acts in view of an aim, on the basis of a project. So, animal behavior is repetitive and mechanical, incapable of memory, set in motion only by natural instincts. Thus it lacks history, or reflective capacity, a critical return to itself.[25] Human behavior, on the other hand, driven by instincts but also by the project, is teleological. That is, it poses itself a goal, returns critically to itself, and thus accumulates memories and on the basis of this collective memory is able to formulate the sense of its own actions. The animal thus has only nature, whereas man has nature and history. Man is a remembering animal who tells stories, invents plots, plays with memories. "Once upon a time. . . ."

History, however, is commonly understood as *historia rerum gestarum*, or the history of things done, of great undertakings. But, and here Nietzsche's originality begins to appear, an excess of history stops one from making history. He introduces here the concept of specific horizon precisely with the aim of bringing out the limits connected to a general vision and thus a general historic sense. According to him, historical consciousness is useful and indeed necessary, but only insofar as it does not block but rather exalts the instinctive part, the will of affirmation and power of human beings. "What is not historical and what is historical are equally necessary for the health of an individual, a people and a civilization."[26]

"While life needs history, at the same time history can become an enemy of the people. Nietzsche distinguishes three types of history which in his view belong to the living: (1) as active being (monumental history); (2) as conserving being (antiquarian history); and (3) as being in need of liberation (critical history).

The first type belongs to those who look at the past only "to draw breath," and then take up the march toward the goal. A goal is not necessarily personal, but concerns a whole people if not all humanity. Faith in humanity is expressed precisely in the need for a monumental history. But here there lurks a contradiction which Nietzsche makes plain. Active men who need the inspiration found in monumental history are also in fatal contradiction with this history as they cannot accept the idea that "the great must be eternal." If they accepted this idea active men would not be able to realize their specific goal. They would no longer have anything to do. Everything would already be said and done. In practice the monumental devalues the present.

Nietzsche gets out of this contradiction by noting that greatness, existing at one time, was at that time at least possible, and thus will again be possible. The antiquarian type of history concerns the being who preserves and worships. History thus becomes history of the great events or monuments, that is, historical history, *historia rerum gestarum*, the history of elites, also history as collective memory of the everyday. He argues:

> What is small, limited, crumbling, decrepit acquires its own decorum and intangibility when the soul of antiquarian man, protective and worshipping, migrates into these things and there prepares himself

a family nest. The history of his city becomes for him the history of himself: it includes the walls, the gates and towers, the municipal artillery, the popular festival like a decorated diary of his youth, finding in all this, himself: his strength, laboriousness, pleasure, judgement, madness and rude habits. He tells himself that here one could live because one can live here. Here one will be able to live, because we are obstinate and things cannot fall apart in the course of one night. Thus with this "we" he looks beyond the ephemeral and wandering individual life and feels himself as the spirit of the house, the race, the city.[27]

Here Nietzsche, with exceptional insight, offers us the essential elements of what is today called cultural anthropology, social history, oral history, or ethnohistory. History is no longer conceived of as the noble sequence of great events, battles, treaties, dynastic marriages, and so on, but rather the cumulative result of the webs, networks of relationship, and material relations of life in which neighbors necessarily, not only at times of great events but in the molecular microhistorical grill of the everyday, enter among themselves. Clearly today's much elaborated conception based on the systematic collection of life histories and oral testimonies is present in Nietzsche's perspective. He seems especially concerned with the degenerative aspects of antiquarian history. These aspects take the upper hand "at the precise moment when the fresh life of the present no longer re-animates and exalts it." Man then becomes prey to an "omnivorous greed for the old." From understanding one then falls into simple erudition, in the dust of bibliographic trifles. The global sense of everydayness is lost. But even when one does not fall into these negative aspects, antiquarian history reaches an insurmountable limit: "it has in fact only the possibility of *conserving* life, not generating it."[28]

Critical history consists in dragging history before a court, questioning and then judging it. But who is the public prosecutor in this court? Whose job is it to decide the judgment and pronounce the sentence? It is enough to pose questions to upset the ancient maxim of Cicero that history is the teacher of life and at the same time to show the fragility of Croce's formula for whom all history is always necessarily contemporary history. Nor, in Nietzsche's view, can one say that justice must judge the past. If this were true, it would be difficult to avoid falling into the metaphysical trap of an ethic conceived of as a metahistorical norm. Most logically, to pronounce the verdict is "life alone, that obscure power, overwhelming, insatiably hungry for itself." Life sentences are never just in the abstract. Life, like nature, is pitiless and cruel as it is never the fruit of a cognitive drive alone, but also of a powerful, uncheckable, instinctive drive.

What services can history render to life?

Each person, each people needs, according to Nietzsche, a certain knowledge of the past. But this knowledge must have as its aim life itself. Hence for Nietzsche life is the end of historical knowledge, history need not be science. Here he takes up again his polemic against history as pure erudition just as he had earlier criticized the philologism of pure philologists. Once more he polemicizes against aestheticizing formalism which in the name of form betrays the essential substance and generates what he calls the weak personality. According to this personality it is inevitable that there should open up an abyss between form and content so that life in its most spontaneous and powerful manifestations appears only as barbarity. The rejection of barbarism becomes for Nietzsche the choice of internality for the German character, or more simply the imitations of foreigners, especially the French. In this imitation he sees a certain sign of inferiority because the imitation always seems negative: "Everything is colorlessness, worn out, ill-copied, gone off, everyone acts according to his whim and not according to a vigorous, thoughtful spirit, but according to the regulations established now by collective hastiness, now by the collective search for ease."[29]

Historical knowledge can thus be an enemy of life in five ways:

- as it generates a contrast between the external and internal
- because it leads to an exaggerated belief in justice
- because it distorts people's instincts
- because it leads to the belief that we are all late fruits and epigones
- because it leads to self-irony and thus to cynicism

All in all, even if he never speaks of free personality, to Nietzsche modern man suffers from a weak personality; he has lost his instinct and is too conscious.

If modern man is weak, can we at least say he is more just compared to men in other times? Does historical objectivity produce only the appearance of justice? Justice would confer the right to great admiration since into this virtue there flow all the others. However, this is not the case as regards modern man. For example, the virtuous historian of the present is certainly not the most just man of his age. Even if nothing human is alien to us, it is only a resounding passivity. In reality, only the superior man can write history, the man of vast experience, as "he who has not lived something greater and higher than all others will not even know how to interpret something great and lofty in the past."[30] The historical sense is thus necessary, but when it domineers uncontrolledly, this same historical sense unroots the future. Why? Because it destroys illusions. And it is interesting to note here how Nietzsche uses the same term and the same category of illusion used by the romantics and Leopardi and Foscolo. Indeed the historical sense carried to extremes paralyzes the vitalist optimism which makes truly innovative human enterprises possible. Nietzsche cites the *Meistersingers* where Hans Sachs says great things "never happen without a little illusion." Nietzsche criticizes the formalistic scientific method in the study of history, just as he condemns the use of philological erudition for its own sake. It fails to give the young a sure orientation and overall vision, but confines itself to making them specialists. In this regard he quotes Goethe, who believed that the sciences should act on the external world only through a higher praxis, that is, not bowing to the undifferentiated needs of the market but respecting the specific requirements of great personalities.

To Nietzsche, our time is characterized contradictorily by great historical education and at the same time a kind of ironic self-awareness, that is, the presentiment that a deep crisis will soon arrive which will mark the end of a historic knowledge understood only as an erudite or academic pastime. In this sense, we are aware, even if not explicitly, that it is senseless to believe that education should thus be based on history as it is today. It is hard to find a great people endowed with such a historical consciousness, as the greatness of a people and its warriors is made up of their capacity to act, while historical consciousness is essentially a medieval *memento mori*. Against it, the modern age cries its *memento vivere*, but humanity's sentiment in the modern age is still intimidated and firmly welded to the Middle Ages' *memento mori*.

The modern age is thus a historical epoch, possessing a kind of primogeniture. But in this regard a grave doubt remains. Modern man in fact is proud, but at the same time self-ironic. Through irony he expresses his unconfessed fear of not managing to save anything of his hopes and youthful strength, that is, his illusions. At times modern man pushes himself beyond his self-irony to cynicism, to the point of accepting the world as it is because it could not in any case be otherwise. Nietzsche quotes Eduard von Hartmann and his famous philosophy of the unconscious. He presents him as the philosopher of unconscious irony. Hartmann announces to our age that it must be just as it is, like an awful ossification, as a "troubled screech of bones." Against this prospect Nietzsche defends youth and puts it on its guard as regards the excesses of the historical sense.

The only hope in fact is that of youth. Nietzsche has faith in the young and in their name expresses a protest against historical education. He asks that man first of all learn to live and use history only in the service of life apprehended. He especially deplores the German concept of culture, false and superficial, as it tolerates the contradiction between life and knowledge. For this reason the aim of the German education of youth is not in fact the free cultivated man but the scholar, or the man of science, setting himself aside from life so as to recognize it in the clearest possible way: "the philistine-aesthete of culture, the knowing, up-to-date gossip who discourses on the state, the church and art, the sensor for a thousand kinds of sensations, the insatiable gut who nonetheless is ignorant of what a real hunger and thirst are."[31] The young must instead understand that culture can also be something different than a decoration to life. As the ancient Greeks knew, culture can present itself as *Bildung*—the real education of man.

But what escapes Nietzsche is the revolutionary character

of tradition. Antiquarian history is not only a quiet, resigned veneration of the past. How can one form, elaborate, and transmit a true *Bildung* without taking up contact and discovering the seeds of the future that tradition bears, hidden from the superficial glance which mistakes the new for the different—seeds within the deep folds of its shady essence? The "omnivorous hunger for the old," in the very tension of its desire, is a movement toward the new. Tradition is not only traditional, nor does it resolve everything in the Weberian "authority of the eternal yesterday." As Benjamin intuited, there remains the presentiment that one day we shall have to go back to pick up something forgotten,[32] that there is a forgotten future near us.

TIME AS DURATION AND AS MECHANICAL UNIT

Everyone knows and admits that tradition is "handed down" (*traditur*) and that in this sense it is the determining factor of stability and psychological assurance. But it is also a whole composed of knowledges, a patrimony of survival techniques. Experience has been summarily devalued in the name of a rootless progress, but today we are aware that there is no authentic innovation, no bearer of new meanings of human emancipation, save in a positive relation with origins. A decisive fact has escaped—that in tradition values yet to be realized are guarded, the historical promises to keep, the values which have mortgaged the future and which alone can give it a human meaning, an intersubjective sense. Making a myth of, and at the same time mummifying the past-tradition, in the framework of the general problem of time in its three basic dimensions, have been most persuasively conceived of thus:

> The reciprocal relation between temporal moment may . . . be broken up, as a result of prevarication centered either on the past or the present or future. In the first case instead of a recapturing of the past one has a claim for repetition. Knowledge of the past becomes a mythical, solidarist projection, with a disposition rejective of any novelty. In the second case, there is instead a claim for unconditional control

of movement. The analysis of the structure is carried out as a function of the mechanical integration of the elements of novelty in the process. ... Finally, if prevarication is undertaken from the viewpoint of the future, the knowledge of consciousness becomes a *utopian* illusion, evasive as regards the concrete tasks and the open possibilities of the present.... For human consciousness there is not provided... what might be called *full temporality*. Historical temporality is, rather, revealed as discontinuous and full of breaks.[33]

Enlightenment progressivism, which still, for example, emerges in the decisions of modern urban planners, of the modernizers and eviscerators[34] has no awareness that tradition is also constant reinterpretation, life taken up and concentrated, value and symbol. Woe to whoever has no sense of roots. His progress becomes senseless. It is resolved into pure, crazy, neurotic agitation, denied the remembering force of its origin.

The subjectivist limit of Bergsonian *durée* is clear here. The individual consciousness is its field of manifestation. It celebrates its epiphany there. Emotions, thoughts, states of mind, memories and nostalgias, fearful waiting, are clustered in it in a progressive and irreversible series. This organization of the data of consciousness is the work of time; indeed, it is time itself. But individual consciousnesses are different in their rhythms, in the intensity of their impulses, in the gamut of their vital tensions, just as they are asynchronic, nonsimultaneous, and dissonant. Time scanned by watches is thus a pure artifice of social convention. It does not concern time effectively lived. Duration as deep, vital, qualitative experience, and mathematical representation of time as a sequence of exactly calculated, quantitatively measured intervals do not correspond. The attempt to capture *durée* in a mesh of collectively established hours and minutes does not make any sense.

However, Bergson is aware of the risk of the fragmentation of *durée* into a myriad of consciousnesses which fall into indistinction and irrelevance. Thus he postulates a single time which crosses from one *durée* to another:

All human consciousnesses are of the same nature, perceive in the same way, walk at the same pace and live in the same duration. Now, nothing stops us from imagining as many human consciousnesses as

> we like, spread at a distance across the whole of the universe, but close enough to each other so that two consecutive ones selected by chance have in common the extreme edge of the field of their exterior experience. Each of these two exterior experiences is a participant in the duration of each of the two consciousnesses. And as the two consciousnesses have the same rhythm of duration, so it must also be for the two experiences.[35]

The reasoning seems plausible, but Bergson probably fears its psychological orientation, tendentially an introspective one. However, his intention is always that of embracing in a unitary, dynamic, essentially open system the whole of the phenomena of the world: those of the psychological individual sphere, but also those of the external natural world, the humanistic and social sciences, and the physical sciences. His field of intervention, the object of his meditation, is the universe; he no longer recognizes fences between human and nonhuman experience. He continues:

> The two experiences [of the two consciousnesses] have a common part. . . . [T]hey join in a single experience, taking place in a single duration and can be considered at will as belonging to one or other of the two consciousnesses. *As the same reasoning may be repeated from any point to the nearest point*, one and the same duration is collecting along its path the events of the totality of the natural world and in the end we can eliminate the human consciousnesses we had placed in the various points—like an equal number of connections—for the movement of our thought: and there will only be the impersonal time in which all things will pass by.[36]

Just so, all things—consciousness, factual empirical data, culture and nature, psychological world and biological/physical world—are united in the great impersonal, irreversible sea of time. Bergson, as it were, lends Prigogine a hand. But the doubt persists whether the process of desubjectivization in his philosophy is logically complete and does not rather concern a true anticipation made by genius. Bergson's thought has been taken in and interpreted probably beyond what the author intended. Bergson is being recovered by divorcing him from the tradition of introspective philosophy which on all the evidence conditioned him. One tries to understand how he was capable of anticipating the work of science, specifically the way in which scientific research develops today, trying and trying again with the utmost problematic openness.

A METAPHYSICS FOR SCIENCE

Bergson's critique of science should rightly be related to the historical moment. He was not a critic of science as such, but of positivistic, dogmatic science, of the old science still subaltern and guided by Cartesian and positivistic premises, by solid mechanics. Today, rather, one perceives in Bergson the forerunner of fluid thermodynamics. He is thereby absolved of the hasty accusation of having submitted to the fascination of the indistinct and the vague taken as values in themselves. Time as complex duration would mark, on the contrary, the point of transition if not the bridge uniting the new sciences and philosophy, the sciences of nature and of man. The illustration of interior time as complex duration, as, in fact, not quantifiable nor to be made substance, anticipates (or at least does not contradict) the idea of irreversible time which gives birth to order through fluctuation, and prepares the possibility of realizing a similar operation also as regards space. Creative evolution is thus about to be freed from its subjectivist orientation to be presented as scientific metaphysics, open as such to physics, chemistry, and the "impersonal processes of nature."[37]

Bergson's construct is essentially dynamic. The spirit cannot live off income. The past is a capital to be used. So used it provides for the future; it is an instrument for ever-new creations. It is as though one were reading Schumpeter's theory of economic development in which the capitalist-demiurge does not await the requests of the market but foresees and creates them. Moreover, this activism is perfectly in line with the Judeo-Christian tradition Bergson felt he belonged to. According to the Jewish religious tradition, he who stops in his spiritual search does not stand still but goes backward. For the Calvinist the certainty of salvation is tied to economic prosperity, achieved by living a holy or methodical and rational life, in which as lived ethic there are welded together vocation and profession (*Beruf*). For Bergson it is the passing of time which is the capital on which one must ceaselessly work. We are then not too far from the practical and sublimated egoism of a Benjamin Franklin: time is memory.

Can one sustain the idea that money is life? That it thus absorbs all life? Certainly one may try to disembody Bergson's thought from the individualism which sustains and justifies it, but it is hard to close one's eyes to the dithyramb Bergson unleashes in favor of the man of action:

> By what sign do we recognize the man of action, generally; he who leaves his mark on the events with which he mingles his destiny? Not, perhaps, by the fact that he embraces a more or less longterm vision with an instantaneous one? The larger the portion of the past he possesses in his present, the heavier is the mass he pushes into the future to put pressure on the eventualities being prepared; his action, similar to an arrow's, shoots with the more force forward the more his representation tended backwards.[38]

We are not dealing with the purloined accents of Carlyle's "heroes" who make history or Emerson's "super-souls" who dominate it as sovereigns and manage it from their gentry-style country cottages, but nor are we so far distant. It is the Hero as it might be understood by the so-bourgeois Third Republic: hyperactive, businesslike, hedonistic, and spendthrift. The spiritualism of the duration *desinit in piscem*.

THE TRANSFIGURATION AND TEMPORALIZATION OF MATTER

Bergson is careful not to fall into the shopkeeping spirit of the philosophers and scientists who construct externalizing typologies of vital phenomena and mistake philosophical reflection for the meticulous accounting of an efficient import-export firm, careful not to let the shelves empty of the most requested items. For Bergson to understand life does not mean cataloging its products but rather understanding the very process of production, life in its making, the warm, fluid moment of the productive act. This means letting go the ambition of discovering and a priori establishing an end for life, a kind of preexisting or hypostasized model which would end up by taking it for granted, opposing rationality and methodical reflection to its distinctive, founding characteristics which for Bergson are creation and development. All this can quite well adapt itself to the spirit, the creative, dynamic thrust which defines the human being, that is, to the consciousness of the individual, generally forgetful of Schopenhauer's sarcasm, who on hearing the word *spirit* used to ask with amusement "And who is this young man?"

What does matter represent as regards consciousness and spirit? With remarkable logical sharpness, even though at the expense of some inconsequence, Bergson draws consciousness from external reality on the basis of phenomenological analysis. Consciousness does not move first of all from the ego, but from the external world, with which it is mingled; and only gradually does it manage to detach the body and we ourselves from that world. Matter then stops being presented as the formless, magmatic original chaos. A whole is made of condensed, defined, and stored images, from out of conscious perception. The spirit takes possession of these images through memory, by founding past and present. Thus matter is transformed into act, from the brute fact that it was. It becomes movement and process, allies itself, and indeed becomes an essential part of the creative process of the spirit. This transfigured matter, in some way spiritualized, is what it is only by virtue of the creative process, which in turn can unfold only in time and indeed is identified with time.

Here one sees the nexus between Bergsonism and the most recent currents in the biological, physical, and chemical sciences. It is time which in its passage allows life to produce the forms in which it explains itself and is the memory, that is, the remembering knowledge of what has been, which smooths the path and allows the creation of the future. As Bergson makes clear above all in *Matière et mémoire*, time, retranslated from the internality of consciousness to the externality of geometric and mechanical nature, ends up by identifying itself with space. And it is in space that time can finally display its creative fullness. The spirit, through the passage into matter, not only fluidifies and dynamicizes the latter but manages to reappropriate itself more completely as spirit.

For Dilthey, as for Bergson, the "empirical course of psychic life rests on processes." And indeed, "each of our states has a beginning in time, changes in it and again disappears in it."[39] Aside from temporal processuality, a second point of convergence between Dilthey and Bergson is the hard criticism Bergson levels against the mechanical mathematization scientists conduct on time, a critique Dilthey takes for granted, whereas he is concerned to establish what it is in the constant, ceaseless flow of time that resists and emerges as a structural datum, a meaningful interconnection. The third convergence with Bergson lies in the belief that the individual consciousness is the privileged site of this resistance to the temporal flux, the theater, as it were, in which time appears as real life and emerges as meaning, congealing and concretizing itself as lived time, so as finally to distance itself from the fullness of Bergson's vital *elan*; generous and omnicomprehensive certainly, but also indistinct and confused. Dilthey's absorbing interest concerns the possibility of enriching the present, widening its borders, drawing from it all its potential expressive and effective richness.

I have indicated here the memory which recalls deep experience. These are the bricks from which one builds the personality—highly individualized, unique, unrepeatable, irreducible to another. Memory does not recall *en bloc*. If one remembered everything one would cancel every trace, giving rise to a shattering parasitism. Memory filters and selects. How? On the basis of what criteria? And how can life already lived, experienced, and empirically worn through be recalled to consciousness? Be truly reactivated? We are speaking of an effective human experience. It is not a story, nor a *mythos*. It is life really lived. Memory every day performs the miracle of the resurrection. Dilthey calls this miracle *Erlebnis*. It is a reliving which cannot be confused either with the literary or the historical tale. It is an experience which bites deep into the soul, which returns to give it the sense of what has been and so what is, and what can still be in the framework of a meaningfulness which transcends it and creates it as an essentially supraper-sonal, metaindividual value. It is no longer reducible to the purely private environment. But the logic of this mediation remains mysterious. How the common *Erlebnis* or the simple *Nacherleben*, reliving, achieve the depth of a real *Erfahrung*, of an experience of the new and unforeseen, is a still-open question.

Recent attempts to speak of time in terms of a tale to interpret, moving from literary criticism rightly so-called with the aid of an interpretative phenomenology, are certainly praiseworthy, but it is doubtful they can clarify the question on a critical plane. It is true that a tale is none other than the expression of a sequence of cases, of events unrolling over the course of time, but it is still less adventurous to hope to arrive at the heart of time as problem of the lived through, however acute an analysis of the intrigues which support the invented tale. Certainly the brilliant comparison of Augustine's *Confessions* and Aristotle's *Poetics* helps to understand the contrast between literary intrigue and the Augustinian *distentio animi* only then to clarify their basic convergence.

In Aristotle we are dealing with the victory of concord over discord, while Augustine groans and suffers under the weight of internal laceration. But that this harsh comparison is a purely intellectual undertaking is clearly and worthily recognized by the person proposing it: "It goes without saying that it is I, the reader of Augustine and Aristotle, who establishes this relation between a live experience in which discord tears concord asunder, and an eminently verbal activity where concordance remedies discordance."[40]

LITERARY STORY AND HISTORICAL STORY

Ricoeur's intention is inspired and explained by a higher ambition. Starting from the invented story, where the plot involves an encounter-clash between characters caught up in a temporal sequence of events tied to the deadlines of a specific chronology, Ricoeur does not delay in revealing his true objective. This is the other type or genre of story, the historical one, where the rigor of causal reasoning or the analysis of the conditioning matrix and the critique of sources and study of available materials create ties which the story of invention does not encounter. The two types of story, however, seem closely linked. This link depends on the simple fact that both the story-teller and the historian are ultimately bound to narrative: they cannot escape the narrative logic of a situation which develops in time.

> Two convictions of equal strength are at the origin of this present study. The first says today what links the narrative character of history to the survival of a particular form of history, the narrative, is a lost cause. In this regard, *my argument regarding the ultimately narrative character of history is not in any way confused with the defense of narrative history*. My second conviction is that if history broke off every link with the *basic competence we have to follow a story* and with the cognitive operations of narrative understanding... it would lose its distinctive character in the complex of the social sciences: it would cease to be historical.[41]

In other words, if history renounced being narrative—which must not be confused with the defense of old-style artistic history, purely literary and hortatory—renounced, in other words, the characters, development of situations, and chronology, it would renounce itself. History would give up being history and "resolve itself into sociology."

The notion of sociology entertained here is fairly crude and unable to stand up to a serious critical examination. It is indeed surprising that a writer seriously interested in removing the Manichean gap between artistico-literary narrative and historical narrative should then be so backward as to continue ingeniously to postulate dogmatic fences between specific sciences of culture and society, such as sociology and history.

However, let us set aside this subject which would lend itself to a subtle logical and epistemological discourse. The fact is that the point of convergence between literary and historical story is their common interest in temporality. Even recent historiographic schools (Ricoeur's explicit reference is to the *Annales* school, but the book duly examines also the representative authors of English and American historiography) who have tried to escape the logic of narrative of the intuitive-logical kind, cannot avoid concerning themselves with time. They will no longer put the great events of history in the front line but they will mainly be concerned with everyday economic and social facts, the evolution of institutions and mentalities. They will, however, not be able to avoid framing their analyses in the temporal dimension. Ricoeur argues that if they lose sight of the temporal dimension they leave the field of methodological options; they lose sight of history itself; they are no longer historians but something else. The critical demands Ricoeur directs against Braudel in this instance, taken here as a personal polemical reason, are particularly cutting and seem to leave no options.

> If the discovery of the long duration did not lead back to the event according to one or other of these three modalities (the event as residue of every explanation; as dissonance between different explanatory structures; as life and death of structures) it would run the risk of tearing historical time away from the living dialectic of past present and future. A long time can be a time without a present, and also without a past and a future: but it is then no longer an historical time, and long duration reduces human time only to the time of nature. ... The fact is that the discovery of long duration may express the *oblivion* of human time, which always requires search for the present. ... This disastrous conclusion can be avoided only if one preserves an *analogy* between the time of individuals and that of civilization: an analogy of growth and decline, creation and death. An analogy of destiny.[42]

So, for Ricoeur, in narrating, either in the artistic-literary or historico-documentary modes, the same operation is always being performed. As it were, *time is humanized.* When we tell a story, any story, true or invented, we always establish a necessary relation. The story achieves its full meaning when it becomes a condition of temporal experience. Up to this point Ricoeur's argument is clear. What spoils it is a rather serious internal contradiction. Ricoeur still presupposes the existence of a time *a parte rei,* according to the canons of an ingenious or paleopositivistic realistic conception, whose insurmountable limits he wishes to stress. Further, he does not seem aware, perhaps for the simple lack of concrete experience as literary narrator, that the logic or structural congruity of the invented story may in fact be much more binding than that of the doings of real characters.

The life histories collected by Oscar Lewis prove this beyond any reasonable doubt. It is enough to make a rapid comparison between Fernanda, one of the protagonists of *La Vida*, and Moll Flanders, heroine of Defoe's novel. Moll ranges herself against the social, it has been convincingly argued, and knows she is a delinquent, transgressing the normal codes of behavior, and it is this, apart from her aspiring to the status of a rich lady, which obliges her to play the game of appearances in her relations with others. Fernanda has no need to hide herself. She is first of all herself. Yet the narrative structure of Defoe's novel is similar to Fernanda's story, and stresses the contrast between past and present. Aside from these important analogies, one should note a remark which seems to invert the relation between tale of invention and historical account as laid out by Ricoeur: "Defoe wishes above all to be a realist and provide a picture of everyday life by describing places and common people. It is typical of Defoe that he has the ability to describe scenes laden with irony, but not that of creating a novel coherent in all its parts. In this sense, one could argue that *La Vida* appears as a more coherent text than *Moll Flanders*."[43] In other words, the novel is more historical than the historical account or the scientific autobiography.

Ricoeur would respond to this observation with the concept of the refiguration of time. The reply is rather fully worked out and deserves extensive quotation:

There is . . . an aspect whereby fiction imitates the historical account. To tell something means in fact to relate it as if it had happened. A clear indication of how this *as if* is essential to the efficacy of the fictional account is provided by the grammatical structure. It is a fact that tales are told with their tenses in the past. The "once upon a time" indicates in narration the "entry-into-the-story". . . . These tenses are also those of memory and history. One cannot just be dealing with a contingent relation. Is this not perhaps why, when the facts of a fiction are narrated, they are taken as quasi-past? For whom, though? The reply I suggest is this: they are such by way of the *narrative voice* which we can here consider as coinciding with the author involved, thereby involving a disguise of the real author. . . .

This interpretation of the almost fictional character of the historical past. Thanks to its almost historical nature the fiction can exercise a liberating function regarding the hidden possibilities of the effective past. What "might have happened," which for Aristotle is the object of poetry as against history, creates a fusion between the potentialities of the "real" past and the "unreal" potentials of the pure fiction.[44]

It would be extreme to reject the idea that the empathetic consideration of literary texts might be useful in the investigation of temporality as a meaningful dimension of the human presence in the world, to use a traditional formula which still retains a certain Heideggerian assonance which would also not displease Ricoeur. However, rather than making perilous comparisons, referral to genuine literary texts can aid us in understanding if not rigorously conceptualizing temporal experience. A good example of this is provided by Maurice Merleau-Ponty, who uses a passage from Proust to suggest that memory, far from presenting itself only as the founding moment of the past according to common sense, really points to the not always successful (and thus dramatic) attempt to "reopen time starting from the implications of the present." Our body "as it is our permanent means for adopting attitudes and thus fabricating pseudo-presents, is the means of our communication both with time and space."[45] Memory and body are thus two essential means for our perception of temporality. It is scarcely necessary to add that they are at the same time both subject to temporality and its usury, and that they are little by little decomposed, almost emptied from within, entropically, to the point of exhaustion and final dissolution.

The general plan of Ricoeur's argument is different. The critical demands he makes of the "new history" and especially the *Annales* school—which can be called new only in a manner of speaking, founded over fifty years ago by Marc Bloch and Lucien Febvre and until recently authoritatively led by Braudel—hinge on the ancient conception of man as *animal confabulans*. They are explicitly aimed against the quantitative scientification of the historical discourse. In this sense they are a worthy attempt to restore the vetero-humanist attitude. But their weakness can be seen in a curious timidity. Ricoeur criticizes the new history in that it burns its bridges with the old historicism but then stops halfway, and cannot grasp the need for a historicism renewed in a broader perspective—a need both theoretical and political. The comparison he sets up with the invented texts should at all events be attempted with the biographical approach and life histories. But Ricoeur prudently retreats to the literary account and rests his whole critical position, in the positive and negative sense, on the analogy between the fictitious literary account and the historical account. But the analogy remains essentially external. The convergence is weak. He confines himself to showing that the two types of account necessarily move in the dimension of temporality. *A la bonne heure*![46]

Postponing the elaboration of this subject to another occasion, a subject whose importance certainly must not be minimized, it suffices here to stress that the use one might make of literary experience is much more productive. The reactivating of past experience, however, seems to reach a depth that a purely intellectualist operation which tends to establish analogies between literary and historical account does not. The question involves not simply the essential importance of temporality, which can in no way be passed over in silence. What is at stake here are the terms in which temporality is lived and relived. Memory is never the simple photographic reproduction of past experience. In memory the past-as-past becomes the present past, relived not only as past but with all the potentialities already present in the past-as-past but now, in the lived and now relived past, presented as seeds of the future.

NOTES

1. F. Ferrarotti, "The Paradox of the Sacred" (English translation), *International Journal of Sociology* 14, no. 2 (Summer 1984).
2. H. de Montherlant, *Carnets* (Paris: Gallimard, 1957), p. 189.
3. Among more recent studies on this, see Jeremy Campbell, *Winston Churchill's Afternoon Nap. A Wide-awake Inquiry into the Nature of Time* (New York: Simon & Schuster, 1987). Despite the rather

general title, the book deals almost exclusively with the biological meaning of time.

4. For a preliminary treatment, see my "Dato pragmatico e dato problematico," in *Lineamenti di sociologia* (Naples: Liguori, 1976); and *Quaderni di Sociologia* 1, 1951.

5. See my "Osservazioni preliminari sul rapporto filosofia-sociologia," *Paradigmi* 13 (Spring 1987), from which I repeat the essential passages.

6. See the iconoclastic remarks of Pareto not only in the *Tractatus* but even more in the minor works often, at times, connected to events of the moment.

7. See my *Trattato di sociologia* (Turin: UTET, 1968).

8. Among the numerous contributions by L. Geymonat, of particular interest from our point of view, are the following: *Scienza e realismo* (Milan: Feltrinelli, 1977); *Lineamenti di filosofia della scienza* (Milan: Móndadori, 1985); and in collaboration with Giulio Giorello, *Le ragioni della scienza* (Bari: Laterza, 1986).

9. G. Friedmann, "L'homme et le milieu naturel" in *Ou va le travail humain* (Paris: Gallimard, 1950), p. 27. (First published in *Hommage à Marc Block–Annales d'histoire sociale*, 1945).

10. Published in Paris by A. Colin, 1942.

11. Published in Westport, CT by Greenwood Press, 1985. See esp. pp. 50–54.

12. Lewis Mumford, *Technics and Civilization* (London: Allen and Unwin, 1946), p. 14.

13. Friedmann, *Où va le travail humain*, p. 63.

14. Published in Paris by Comunità, 1942.

15. Friedmann, *Où va le travail humain*, pp. 59–90.

16. Friedmann, *Où va le travail humain*, p. 62.

17. N. Elias, *Saggio sul tempo* (Bologna: Il Mulino, 1986), pp. 9–10.

18. As regards this basic dichotomy on which Heidegger's thought (among others) hinges, I refer the reader to my *Storia e storia di vita*, esp. chap. 1, sec. 2.

19. D. Luzzatti, "Borges, Bachelard e il tempo," *L'uomo, un segno* (Jan.–Apr. 1982): 4.

20. Robert Lowell, "Near the Unbalanced Aquarium," *New York Review of Books*, 12 Mar. 1987, p. 10.

21. Jorge Luis Borges, *Finzioni*, Italian translation (Milan: Móndadori, 1988), p. 95.

22. W. Benjamin, *Reflections*, English translation (New York: Harcourt Brace Jovanovich, 178), p. 30.

23. A. Gide, *Ainsi soit-il ou Les jeux sont faits* (Paris: Gallimard, 1952), pp. 34–35.

24. R. Bodei, "Le malattie della tradizione," in *Walter Benjamin* (Rome: Editori Riuniti, 1983), p. 215.

25. For information on the current debate about the relation between biology and the social sciences, see Arthur L. Caplan, ed., *The Sociobiological Debate* (New York: Harper & Row, 1985); and my preface to H. Spencer, *Principi di sociologia*, 2d ed. (Turin: UTET, 1987).

26. F. Nietzsche, *Unzeitgemässe Betrachtungen*, Italian translation, from the critically established text by S. Gianietta and M. Montinari (Turin: Einaudi, 1981), p. 86.

27. Ibid., p. 98.

28. Ibid., p. 101.

29. Ibid., p. 107.

30. Ibid., p. 124.

31. Ibid., pp. 153–54.

32. See Benjamin, *Reflections*.

33. F. Totaro, *Produzione del senso* (Milan: Vita e pensiero, 1979), p. 63.

34. In the United States every skyscraper has a map which will make its demolition easier once the investment has paid off; the "eviscerations" in Rome in fascist times are further evidence.

35. H. Bergson, *Durée et simultanéité* (Paris: Alcan, 1926), p. 58.

36. Ibid., pp. 58–59, emphasis added.

37. For a serious attempt in this direction, see P. A. Rovatti, "La 'nuance,' Note sulla metafisica di Bergson," *Aut Aut* 204 (1984): 81–94.

38. H. Bergson, *lénergie spirituelle* (Paris: Alcan, 1908), p. 16.

39. W. Dilthey, *Critica della ragione storica,* Italian translation (Turin: Einaudi, 1954), p. 61.

40. Paul Ricoeur, *Temps et récit* (Paris: Seuil, 1983), 1:55. But temporal duality, between the time of history and the time of the story, is already considered by Gérard Genette, *Figures* (Paris: Seuil, 1972), 3:77ff. Genette in turn refers to the distinction between *Erzählzeit* and *erzählte Zeit* worked out by Gunther Müller, *Festschrift für Kluckhorn* (1948), now in *Morphologische Poetik* (Tübingen: Mohr, 1968).

41. P. Ricoeur, *Temps et récit*, p. 133.

42. Ibid., p. 312.

43. L. Ferrarotti, *Oscar Lewis biografo della poverta'* (Rome–Bari: Laterza, 1986), p. 121.

44. P. Ricoeur, "Il tempo raccontato," *Aut Aut* 216 (Nov.–Dec. 1986): 37–38.

45. M. Merleau-Ponty, *Phénoménologie de la perception* (Paris: Gallimard, 1945), p. 211.

46. The criticisms which the more recent tendencies of the *Annales* have attracted are much more incisive: see F. Dosse, "L'histoire en miettes," *Espace-Temps* 29 (1985): 47–60. For a full but uneven review, see G. Gemelli, "Le 'Annales' nel secondo dopoguerra: un paradigma?" in *La storiografia contemporanea—Indirizzi e problemi* (Milan: Il Saggiatore, 1987), pp. 5–38.

2

The Context

MEMORY AND IDENTITY

Memory is the memory of the experienced. Memory is thus relived time. But this reliving, which has its basic moment in the consciousness of the individual, is not a phenomenon exhausted in the individual or which takes place in a social void. Memory awakens in a precise context, in the relative stability of a situation which spans the biographical spread of the subject. Once more the truth is confirmed that the problems of the individual are not an individual question. Men as individuals make history but history does not belong to individuals; it crosses over the private individual sphere, even if it cannot dispense with it. It is the extrasubjective web which the subjects need to recognize and in which they identify themselves in the dialectical relation between various levels of historical time and the plane of existential time. From this viewpoint, it seems one must perceive an essential ambiguity in Dilthey's conception of *Erlebnis*. This ambiguity is shown in some internal tensions, or contradictions, which we have indicated above and which are worthy of careful consideration.

The *Erleben* seems to present itself as a self-sufficient process in that it is immediate: man thus lives in the individual specific temporal moment. His immediate lived experience is the nourishment, as it were, of his awareness; but then, it is fair to ask, how can one move from the *Erleben* to *Verstehen*, from the lived to understanding?

Man is essentially a historical animal; his historicity is his peculiar nature, not in general but in a specific sense, in that man is tied *in* and *by* his context. When speaking of man, then, one can never speak of him in general, but rather of man *in a situation*. Though tied or conditioned or partially determined by the situation he is in, however, it seems that man has a possibility of choice, if only, in Dilthey's terms, in order to make comparisons with other historical times, to feel himself part of a single universal history. It is uncertain how this is possible.

By their nature, *Erleben* and consciousness coincide; *Erleben* by itself allows no empirical investigation or intersubjective standardized analysis. But if this is so, how can one get away from subjectivist immediacy? And, on the other hand, how can one then attempt critical establishment of sciences of the spirit as constituting a cognitive connection tending to an objective knowledge of the concatenation of *Erlebnisse* in the historical and social world of man?

It seems clear that this concatenation requires getting away from the individual *Erlebnisse*. In the theoretical impossibility of finding a noncontradictory way out regarding the overall design of Dilthey's thought (however praiseworthy his nondogmatic problematically open orientation), his concept of historicity ends necessarily by relating to an impoverished and mutilated version of historicity. His thought meets here with a limit I believe insurmountable. Essentially meaningful remembrance, what Dilthey defines and refers to as *Erlebnis*, does not just recall past experience as such, that is, the already lived and forever concluded, but also the potentialities still unexpressed in experience. This experience recalls and makes the past-future live, the seeds of the future hidden in past experience, the presentiments of what will come, a sense which goes beyond the specific, determinate experience—what is presented to the individual in the enigmatic form of an obscure sense of destiny.

In this sense, memory is the child of time but also the moment of victory over time. It is also its father. This paradox was long grasped and then often repeated in German culture. For Otto

Weininger, mindful of Dilthey's lesson, the man who manages to descend into his interior depths, which is the mark of his greatness, is he who scores a victory over time. Memory, in which man grasps himself by removing himself from the flux of individual discrete events, is the complete victory over time. For Weininger the genius is thus the truly atemporal man. What does this mean? If man essentially coincides with his own historicity, atemporal man would be nonexistent man. It is not a reductio ad absurdum, but a logically necessary corollary. But with this unhappy formula, however romantically suggestive, Weininger simply wants to point to the man of genius, or the man able through memory to gather and connect in an overall picture his individual, separate, discontinuous experiences. "The individual temporal moments of his life are not given by the memory of the gifted man as discrete points. . . . Self-analysis, rather, means the individual lived experiences appear as joined together in a wholly enigmatic way; events do not succeed one another like a watch's ticking, but run together in a unitary flow, in which there is no discontinuity. This authentic continuity may assure man he is alive."[1]

Memory as an instrument of self-formation plays a basic role also in F. Amiel's *Frammenti di un diario intimo*. But it is probably in Wittgenstein's *Secret Diaries*, that memory, rather than being the mere tool of an essentially narcissistic everyday exercise or of idle internal meaningless probing, transforms itself into the tormenting and at the same time decisive moment of self-identification and the constitution of personality. In the introduction, Aldo Gargani explicitly shows that the internal search for oneself leads to an overall vision of the experiences of life. He sees this need for internal reflection in the itinerary of Wittgenstein's philosophical production, as

> for him, too, descent into himself, the gathering of himself with unflinching courage, is a tormenting process at the limit of which there flowers the penetrating representation [*die übersichtliche Darstellung*] which is the essential aim of his philosophical work. Search for connections, the intermediate passages among the dispersed data of our life in an overview [*Ueberblick*] is the element in which the very function of Wittgenstein's philosophy is defined.[2]

CONVERSION

This revealing of oneself undoubtedly oscillates between a moment of high interiority, which touches on mystical rapture, and aesthetic ecstasy. It amounts to a self-revelation which can have the strength and weight of a real conversion, a radical change of direction, the discovery of one's real self and of one's own road and profound calling. Not by chance in Augustine's writings do we find extraordinary testimonies to this effect: "Do not try to leave yourself, return into yourself, truth lives within man and if you find your nature changeable, transcend yourself. But remember that when you transcend yourself, you transcend your rational soul. So, tend towards where the light of reason burns."[3] And again, in conversation with Monica in the quiet of Ostia by the Tiber:

> Our discussion arrived at the conclusion that carnal pleasure, however great, however great the splendor of the body it emanates from, far from being compared to the joy of (eternal) life, does not even deserve mention. And with the more burning affection, lifting ourselves towards what is ever the same, we passed through little by little all bodily things and heaven itself, where sun, moon and stars send out their light on the earth. And again we rose internally, thinking, discussing and admiring your works; and we arrived at our spirits and transcended them to reach the region of inexhaustible fertility where you fed Israel eternally on the pasture of truth, where life is wisdom and by whose work all these things are, and those that were and will be, while it is not completed but is as it ever was and ever shall be. "Was" and "shall be" are not in it, but only "is": for it is eternal, and "was" and "shall be" are not eternal. So while we spoke and strived with keen desire to that—there! *we reached it for an instant, no more than a heartbeat, and we sighed.*[4]

Perhaps here more than in the Milan garden when the voice ordered the doubting Augustine "*Tolle et lege*," there is present the reliving of the magic moment when the spark of self-recognition is struck, the sense of a certainty and at the same time a fullness in which pure intellectual knowledge is mingled with a deep emotional impulse. Centuries later, another Augustinian, Martin Luther, had to express himself similarly of the link, glimpsed in a flash but probably after long, tormented meditations, between faith and salvation, in the famous experience of the tower.[5] Analogous must have seemed the "mystical night" for Pascal. But no *metanoia* is ever a lightning flash from a clear sky. Conversions do not drop from the clouds. They are a result, not a beginning. They are in fact prepared by long, intense internal travail which corresponds to the deep dissatisfaction with one's own condition and the general status quo. In a positive sense, in this dissatisfaction there is a very strong aspiration to the different and the new, which is translated into an uncontrollable utopian impulse.

From this viewpoint, the crisis in Augustine's garden, the experience of Luther's tower, and Pascal's mystical night have common characteristics. None of them takes place in a social void. To forget the context means condemning oneself to incomprehension. In Augustine's case, Arthur D. Nock's remarks are helpful:

> Augustine was inflamed by the tale of the conversion and public confession of faith of Vittorinus, one of the body of erudite defenders of paganism in Rome, and again when hearing of the marvellous life and miracles of the hermit Antonius. He reached a state of conviction where he was no longer capable of acting. He had aspired to continence, but it was not yet time. Then came the moment of the crisis in the garden. One day, immersed in the travail of his thoughts, he heard a child repeating "Take down and read, take down and read" (*Tolle et lege, tolle et lege*). Then he took Paul's text which he had been reading, and threw himself on Romans xiii, 13, "Let us walk honestly, as in the day; not in revelling and drunkenness, not in chambering and wantonness, not in strife and jealousy." Such an unexpected voice was, for a pagan as for a Christian, a voice from on high. For Augustine it marked the certainty he had almost but not completely reached; like a long convalescence whose end is sometimes a casual circumstance which finally gives a man at last the sensation of being cured. . . . This is a conversion entrusted in the last resort to the lasting effect of an early impression, of the religious atmosphere in which his mother had surrounded his infancy. Christianity has almost always been for him the religion probably closest to the truth. *Conversion thus grafted itself on to a sentimental background.*[6]

His was a sentimental background, certainly, in the sense that Christian practices in Augustine's case were exemplified in his mother's behavior, but it was also a social background. It is difficult if not impossible to take this extrasubjective background into account on the basis of Dilthey's outlook. Conversion marks a significant arrest in the daily routine; it calls into question past life and promises a different future life. Herein its drama lies, its gamble, suspended between memory and hope as it is. This is also the reason that conversion is a significant moment in a person's life. It is in fact the *Erlebnis* which sets itself off from other vital experiences, typical of the everyday, which follow one another with relatively uniform motion, without disturbing the calm waves of average existence. But it is precisely *discontinuity* which gives meaning to all past experiences, and to all future experiences, which correspond to the positive plan implicit in every conversion or every proposal for a "new life."

In this regard

> it is certainly not the case here of temporal dimensions simply succeeding each other. Rather, this involves experiences where the sense (or mystery) of existence makes itself known ecstatically, rather than an irrelevant continuity of moments. In this sense it seems to me the important fragment on *Erlebnis* of 1907–8 should be read (*Gesammelte Schriften* vi, pp. 313–6). What properly characterizes the *Erlebnis* of the present is defined here as a contemporary presence (*Präsenz*) in it of the past as memory, that is as remembrance of more or less chance events which make up the historical and spiritual essence of man, and of the future as fantasy: that is, as activity which proposes and pursues as aims its own historical and spiritual possibilities (*ibid.*, p. 232). The present in this sense is what Dilthey calls *Erfüllung mit Realität* (a term translatable as realization or assertion of reality, *ibid.*, pp. 72–3, 193–4). This is understood as a temporal confirmation of reality in the meaningfulness of events which discontinuously fill up time, as a stormy superimposition in the everyday of the essential, between what results from temporal repetition and anticipation.[7]

However, there remains a double problem one cannot ignore: First, how can one grasp and ultimately evaluate, if not measure, the distinction between routine and deep experience, that is, significant ones? And, second, how is it possible to render the transition from deep introspective experience, essentially individual-psychological (since one cannot see how supraper-sonal significance can be attained when it exists), to the ex-traindividual social context?

At this point the critical remarks Maurice Halbwachs raised[8] against Bergson's conception of time as vitalistic continuum are again useful. He reveals its attempt, tendentially irrationalist, to enrich a now mortally impoverished human experience, and to conceal its unmendable fracture with a traditional way of life not yet totally commodified. Reality can no longer be masked. This reality is that after two centuries of progress, believed to be irreversible, uninterrupted, and inevitable, we are poor, irremediably poor and stripped, without aura, doubly bound to the utilitarian calculus which dogs us close at hand and scans the stages of our days. Benjamin has grasped with great precision the poverty of meaning of so-called well-off modern societies: "We have become poor. We have given up the human heritage bit by bit, we have often had to pawn it at a hundredth of its value, to receive in advance the small change of the present."[9]

MEMORY, CONTEXT, AND TRADITION

The contagious character of spiritual poverty which seems paradoxically to accompany and manifest itself in the very moment the subject proclaims his total autonomy, has been noted with relation to idealistic historicism, but also to historical materialism.

The historicist believed he was accumulating the wealth of history and experience, but then realized that he too only received the "small-change of the present." The historical materialist rejects the past, on the other hand, as a whole and also his inheritance in a passive and patrimonial sense in order to choose the poverty of contemporary experience as a point of departure. He rejects the *Erlebnisse* for a poor happiness, without cornucopia. He rejects the stable horizon of the historicist present and norm in favor of that exceptional state where the oppressed permanently live. From the past there arrives, with a free and momentary relation, only the charged memory of the future, the citation which flashes in the moment of danger. Solely in this way is there hope of making the meager treasure of present experience bear fruit, of making do with little.[10]

There is no possibility of memory without tradition. One cannot build a world without collective support, the faces and voices of the "world of the fathers." When Ignazio Silone went one day to Israel, before a landscape of arid hills and desert, he said: "But I have already seen this view. It is the sunburned landscape of Marsica where I was born." It was his landscape of the soul, but not his in the subjective sense of the term. Memory reconnected him, backwards, with the sights and landscape of infancy, the long conversations of a pretelevision society with old people. Memory is never a purely individual gift. Memory connects us along the chain of the generations through language, usages and customs, with both collective history and the history of those without history. Halbwachs's example is notably revealing: When I go to London I see at the same time the London my eyes bring me and the London of Dickens.

In other words, there is no memory that is not also an inter- and contextual reconstruction. It is from this point of view that one can accept Horkheimer's harsh critique of Bergson:

Bergson tried to make a philosophy of concrete time: reality, in his view, is not understood as a kind of solid body which lengthens only under the effect of time "the fourth dimension of space"; it is itself a development, a transformation, a change. But this daring attempt which at bottom abstracts from human history, was destined to failure. By asserting a supposed spiritual interiority of the universe, drawn from analogy with internal time, the time lived by the individual, and by inventing the divine *elan vital* conceived as absolute being, Bergson was equally obliged to deny time. His metaphysics, his pantheism is long obsolete, contradicts his conception of the temporal character of reality and annuls it. This contradiction which is moreover found in the gap between language and the method of his philosophy and the ingeniously mythic passages of its contents has from the start bestowed an equivocal aspect on his work. The metaphysics of time does not exist. On the contrary, this is only a beginning contradictory in itself.[11]

We are far from the current revaluation of Bergson, today indicated as the brilliant forerunner of the most advanced physics, of thermodynamics and the transition "from being to becoming" which has found its convinced advocate in I. Prigogine. No doubt Bergson felt he had resolved ancient philosophical questions. Zeno's paradoxes were elegantly clarified and dissolved with his conception of the substance of the universe as unified and immediate moment. The same abhorrence of a vacuum, the fear of the void, seemed to him overcome. To him, it would suffice to think of each thing *sub specie durationis* to immerse oneself increasingly in real, cosmic, eternal duration. We ourselves, says Bergson, become eternal since we become part of that universal duration; we live, move, and have our being in it. Horkheimer limits himself to add submissively, but not so much so as to interrupt the dithyramb of the cosmic order, and we die.[12] In fact, "Bergson the metaphysician makes death disappear. Like the theologian who promises men eternal life, Bergson passes over the fact of death thanks to an exhortation on the eternal reality to which we can all be reunited; he thus shows that his work has the same function as religion and the modern philosophy succeeding him; this would like to console men for how they live down here by making them mirror their supposed eternity."[13]

However crude it may appear, so as to reecho (above all as regards religion) the outdated conceptions which reduce it to a residual and essentially irrelevant phenomenon, destined to undergo the inevitable eclipse of the sacred which will occur in industrialized societies, Horkheimer's critique becomes persuasive and important even today when it addresses the subject of historicity as historical life in the broadest sense.

Reality is not one, nor an eternity. Men suffer and die in solitude, for different reasons. To state that reality is indivisible on the basis of its essence seems to contradict a fact that characterizes history—at least in its form as handed down to us. Men are divided into happy and unhappy, dominant and dominated, healthy and sick. The concepts in which we express this separation, its causes and consequences are undoubtedly formed under the action of a spatializing and classifying understanding, but they depend on historical conditions; their structure is rooted in the psycho-physical situation of the conscious subjects as in the objects marked by history.[14]

OBJECTIVE LIMITATIONS AND THE RHYTHMS OF THE SOCIAL

Here, the importance of givenness, or of objective limitations. Here, the weight of context. There is no absolute experience of time. There is history as the knot of multiple times, of different rhythms, attitudes, and orientations which only field research can bring to light, describe, interpret, explain, predict. The very biographical affair, as it develops and is told in a life history, is not fully comprehensible save within the contextual coordinates which include and variously pass over it, those which led Durkheim to write of the hardness of institutions, of social facts as things, capable of exercizing over the individual a coercive power, and with which the individual must daily come to terms. Without contextualization there is no understanding.[15] The relatively stable parameters of the sociolinguistic and economic environment, and of the political and social era within which an individual life runs its course shrink. The parameters in which he makes his choices, experiences successes and defeats, and arrives at his end narrow. Perception of time by individuals and social groups is in proportion to the context. It varies not only from one age to another in a diachronic sense, but from culture to culture and society to society, even when it is a matter of cultures and societies present and interacting on the synchronic plane. In the first part of this book I pointed out the contrast between the peasant world and industrial society as regards perception of time, the way of measuring and evaluating, organizing, "spending," and "treasuring" it. But within the individual societies, whether they be paleotechnical or industrialized and technically advanced, the perception of time may vary greatly according to social class, level of education, occupation, and the different stages of life. One could reasonably say that societies have their music—particular, irreducible, and not exportable—and that every human society has a deep core which regulates its rhythm, the cadence of its movement, the order of its beats, its aura.[16]

No positive change is possible without bearing this rhythm in mind. It is here that tradition and future meet and assist each other. The failures of modernization in Third World countries must be related to this forgetfulness, of enlightenment origin, as distracted as it is atrocious. I noted above that history is a knot of times in cross-rhythm. The myth of development is the rationalization of Western intervention, the most obvious expression of the Eurocentric incapacity to communicate in humanly meaningful terms with technically backward countries. Development has been flattened out and reduced to simple quantitative expansion. Thus development has been understood and carried out in purely technicoeconomic and financial terms, producing instead of situations of welfare, colossal external debts, the destruction of traditional agricultural practices, urbanization without industrialization, social breakup and human pulverization.

And yet to the average Western mentality, the failure of modernization is inexplicable. It is scarcely necessary to remark that there is no underdevelopment as destiny, as permanent and unmodifiable datum. It is tied to rhythm, to the sense of time, which is specific and different for every culture, and which Western European culture has difficulty in recognizing when it does not coincide with its own, from working times and distribution times to that of payment of rates of interest on debts contracted. Underdevelopment is essentially a gap, a being out of step. As regards the Third World, the error of the West lies in a kind of impatient deafness. It does not know how to listen. It has no patience to listen. It has no ear for others' music. It cannot get away from its own noise. No wonder the otherness of others appears only as territory for conquest, barbarous cacophony, a chance for domination, infidel souls to save at the price of burning their bodies.

THE SENSE OF TIME AND CONTEXTUAL DETERMINANTS

Time, rhythm of development, and context are closely connected. The sense of time is a structural part of the context. It is not a plastic, secondary, psychological appendix. It is its substance. It indicates its consistency. It conditions everyday behavior, economic development, political habits, the quality of life. There is a paradoxical aspect to it which may appear disturbing and which condemns the analysis of temporality to remaining an open book, with the alternative of falling into irrelevant dogmatism. Time in fact passes and stands still, lasts and changes, is immobile and shifting, is ever different and ever the same. These contrasting aspects, or dimensions, interact according to the logic of a complex relational dialectic which awaits exploration. We must content ourselves with illustrating with some crude examples their much-indented profile.

The sense of time among North African peasants shows some special characteristics which appear in sharp contrast with the sense of time of those who live in the urban-industrial environment. There is a large literature on this contrast as I have already indicated. However, the two senses of time are often counterposed in a misleading way. It is true that the African peasant, with his traditionalist mentality, tends to resist novelty and has difficulty in planning the future, especially if the plan involves determinate decisions of middle or long term, and if one has to proceed to a reasoned choice of different goals by establishing an order of priorities between desirable objectives which are mutually incompatible. But this is certainly not an exclusive feature of the African peasant, sufficient to condemn him to economic, social, and cultural inferiority. Time in the African peasant society is lived with apparent thoughtlessness, distractedly, as though in reality time were an unlimited and inexhaustible resource. In fact, what is most striking is the absence of hurry. Even if it may sound paradoxical, added to this we should note a classical sense of measurement. The phrase of an African peasant, clearly before liberation from France, is revealing in this regard: "The French behave as though they never had to die." The Greeks and Latins said the same: *medèn ágan, ne quid nimis*.

One must ask oneself if the tendency to waste time is characteristic only of North African peasants. Even in wholly or partially industrialized societies one can find similar behavior. In Palermo, as in other Italian cities, a scheduled business meeting may not occur four or five times in a row—not for tactical or strategic reasons, related to negotiations in progress which are exceptionally tortuous or difficult, but by simple negligence, misunderstanding, or previous delays. In New York the average delay acceptable is not much beyond the conventional fifteen minutes. In Rome it can stretch to an hour or more.

In technically advanced societies, on the other hand, time is a scarce commodity, to be invested with prudence, to increase like a stock, to grow like a treasure, to be bought. These are not vague metaphors, but precise, measurable realities. Programs for the extension of human life, for example, are on the increase in the United States, where the struggle against aging and to buy time is in full swing. Robert Reynolds of the Life Extension Foundation in Florida has no doubts about the positive results of the battle against aging: "The extension of human life will happen whether we want it or not. If you want to make predictions, you could think of a life of a thousand years. I guess that seems unbelievable, but an engineer in the Victorian age would have said the same about TV. Who knows?" Dr. Roy Walford, a pathologist at the University of California, has cut his weight to 60 kilos by reducing his calorie intake and taking vitamins.

> It's undernutrition, not malnutrition. Sufferers from anorexia nervosa can fast till death, but if they are studied over time they are found in excellent health and able to resist infections because they follow a low calorie diet. I am on a life curve which ends at a hundred, but I don't know when I'll drop. One can slow down ageing and extend the life span up to 60%; I think those born at the end of the century will live to 120 years and that there is a very good chance of living to 130. People believe this means being decrepit, but in fact one prolongs youth into middle age, so that the question to ask is: "Do you want to be young longer?" The problem of you Europeans is that you accept your individual destiny more than we do. And this is one of the reasons why innovative culture has shifted towards the Pacific.[17]

The key word here is *acceptance.* The African peasant does not understand efforts to prolong life, the more so as the chronic haste of the Westerner seems to burn him up early and at least prevent him from enjoying life. The African peasant, the farm-worker of Brazil, like, at one time, the Southern Italian peasant accept time; they do not believe they are immortal. They adopt the words of Ecclesiastes without necessarily being conscious of it: "For all things there is a season and a time for everything." Nothing in the behavior of the African peasant betrays a concern that he may be forgotten. Tradition reigns sovereign. The individual as unique subject, unrepeatable and irreducible to others, is not yet born.

However, it is true that the fear of the *homo religiosus* is that of being forgotten: "Father, father, why hast thou forsaken me?" is the anguished cry of Christ on the cross. Augustine echoes this, positively, with the gratitude fitting for the prodigal son returned and redeemed, in book 13 of the *Confessions*: "You did not forget me, though I had forgotten you." The peasant and plantation worker who wait in the evening patiently in line with comrades on the roadside for the clanking bus which will take them home from the cane fields do not fear being forgotten, because they cannot be. They are still part of nature, existentially implanted in it; they follow the rhythms set by the alternation of sun and moon, docilely, that of the seasons and agricultural activities fixed from time immemorial into procedures and actions forever the same. They do not know, and still less feel bound by, the temporal standards of growth.

In this sense the African peasant thinks that the "French," in this case representative of Western European mentality, believe themselves immortal. They work for unlimited economic expansion driven on by anxiety for an ever greater global production and productivity per man hour, their eyes fixed on the stock prices, feverishly concerned to measure the rise of share prices. In these conditions it should not be too surprising that "social time" in its various forms should be strongly affected by the operational needs of the major productive organizations. At least until recently, social life clustered round and modelled itself on the functional needs of the great factories with the same cohesiveness with which in medieval times city life centered on the cathedral and the castle, and was regulated by the chimes of the church bell-towers which according to Schiller's *Die Glocke* called the people to gather in time of ceremony or emergencies, conjured away the storms' lightning, commemorated the dead, and guided the living. With the coming of electronics, informatics, and telematics, the possibility of transmitting data over distance instantly, with virtually no loss of energy, has made possible a related deconcentration of productive units, but the rigidity of collective times has not lessened.[18]

THE SLAVERY OF "COLLECTIVE TIMES," ARTIFICIALLY DETERMINED AND IMPOSED

It has been convincingly remarked that a watch on every wrist and clocks in shops, streets, and cars is not only a product of consumerism, but also the sign of the diffusion of social time. Time is at once a bond and a resource. It is a bond because, following the growing division of labor, our everyday action is found strictly dependent on that of numerous other actors whose times must be taken into account. It is a resource because the same interdependence has forced us to observe the times of other people and make them coincide with our individual ones. In other words it has forced us to organize ourselves, to plan and foresee our actions and initiatives over days, months, and years. A reliable measure of the extent of the economic development of a society or, as it is also said, its modernization, is certainly the respect of punctuality and the quantity of time, occasionally measurable in years, necessary to establish beforehand the positive result of any project. The old protocapitalist saying, "time is money," applied chiefly to economic activities has now spread to the whole of society and concerns the most minute daily activities.[19]

It is a fact that in the conditions of the urban-industrial environment, control of time is a basic tool of power. By regulating time the activities of citizens are controlled. When Caesar took power in Rome, the first thing he thought of was reforming the calendar, starting from Rome's 708th year (46 B.C.). The great monocratic structures did not only reform the calendar. They changed ideologies and epochs by placing themselves at their origin. Thus they legitimated themselves as sources of time. The Italian fascist era also wanted its time. It lasted a little over twenty years. Hitler, who dreamt of a 1000-year *Reich*, had to content himself with an era less than half that of the fascist. The church, with ancient doubts and re-thinkings, did not behave much differently.

The history of Christianity is largely the history of the efforts of the church to impose a Christian time on the world. The calendar promulgated by Gregory XIII in 1582 was adopted by some German Protestant states in 1700, by the Swiss cantons in 1701, by all German Protestants at the suggestion of Frederick of Prussia in 1744, by Great Britain in 1752, Japan in 1873, China in 1911, Russia in 1917, the Orthodox Church, with some exceptions in 1923, and Turkey in 1926. As for the Christian era, it was probably introduced by the shiite Dioniges the Little who was an abbot at Rome and calculated wrongly that the first year of Christianity coincided with the 754th year of Rome. The first year of Christian power was obviously that of the birth of Christ, but Christianity was long divided on the choice of the moment from which that year would start—the circumcision (1 Jan.), incarnation (25 Mar.), or nativity (25 Dec.)?[20]

The control of time is a basic instrument of social and personal control. Already two thousand years ago the rabbinical schools tended to regulate individual and collective behavior through a precise daily timetable of services. In Christianity it is equally well known that much before the Protestant ethic and Calvinistic moral rigor, the Benedictine Rule, traditionally summed up in the motto *ora et labora* (meaning that it embraced both religious practices and daily, as it were, lay activities) established strict timetables of everyday life in all its aspects. Few exceptions were allowed. The monastery must function with the greatest regularity and punctuality independently of the conditions of atmospheric time and climate (the seasons, natural light, or the relative length of days and nights). This heralds the current productive and social organization which has completely broken its relation with meteorological conditions, with what used to be called "natural time." Perhaps an important connection is to be seen here, which links the great monastic orders to the Protestant Reformation and the deontology of the *Beruf* in its double meaning of vocation and profession, based on the methodical use of available time. Sociologists have correctly stressed that the tendency to planetarization emerging in the last centuries has contributed to that synchronization on a world scale to which the French Revolution vainly aspired, and, later the "positive religion of the Supreme Being" with its calendar and philosopher-saints, promoted ardently and in vain by Auguste Comte.[21]

TIME AS A WHOLE MADE UP OF QUALITATIVE MOMENTS

An era is the sure sign that a new star has been born in the firmament of power. It marks the beginning of a new order, the date which divides history into a before and an after. But real power cannot be content with this. It must gradually penetrate, influence, occupy the whole of life. It must dictate not only the train, which obviously must arrive on time. It must impose working hours, the length of the phases of production, the time one enters the factory or office, when one can close up shop, the holiday period, the dates of religious and civil ceremonies. There is only one calendar. In reality, there are many calendars. There are working, scholastic, parliamentary, sporting, judicial, holiday, and traffic calendars. The more technically advanced a society is, the more it restricts free individual choices. Only a relatively primitive society allows wide freedom of movement, save naturally the social control from the community, which may be paralyzing.

In the case of the Algerian peasant, studied years ago and certainly before the liberation of Algeria from French colonial control, we find in Pierre Bourdieu's study total submission to nature first of all, the "submission to the passage of time as it is accentuated by the rhythms of nature."[22] But there is more. All the procedures and techniques of Algerian agricultural work have a clear traditional imprint, surrounded by a halo of religious ritual, so that they acquire a symbolic value and are integrated into a system one might define as mythology in action. As in any *mythos* a contradiction must be brought out here which touches on tragedy. Certainly the peasant loves the soil which sustains him and his family with its fruits. But to gather the fruits of which it is potentially capable, the earth must be violated, in a real sense wounded, ploughed, and worked with the scythe and plow. No wonder that the highest moment of agricultural activity, that of the summer reaping and the harvesting of the crops, is also the nub of the tragedy which takes place between man and nature.

What is striking in this tragic relation, at least to modern Western eyes, is the absence of any concern for productivity and working times. Working timetables do not exist as factors of collective coercion. Certainly time exists, but only as the perception of the passage from one season to another, from day to night—a perception in which the points of reference are not quantitative data but rather qualitative moments, not fixable in a precise mathematical sense, but as visual experiences, sensations, existential memories. Time is not thus a uniform line growing to infinity. Rather, it is a complex of separate and juxtaposed moments, corresponding to as many experiences, so that the week may be indicated with the phrase *es-suq*, from one market to the next, or the lapse of time between one market and another. Thus time is not a unitary flux; rather it is a fragmentary and discontinuous series of experiences closed in on themselves, and meaningful in themselves in the life of a person.

THE ILLUSION OF THE PLANNED FUTURE

It is clear that such an attitude precludes any possibility of planning the future. To plan the future means foreseeing a certain order in actions, working out a table of priorities for initiatives. This involves a notion of linear relatively homogeneous time, logically mastered in the middle and long term. The Algerian peasant, by way of his notion of time as a whole made up of relatively independent experiences, cannot conceive of the future in the real sense, but only what enters the immediate horizon of possibilities already experienced. Bourdieu's concern not to counterpose the Algerian peasant to the technically advanced European schematically can well be understood and shared, and it is true that in the sleepwalking of the everyday the Westerner, like the technically backward Algerian, lives in the immediate and behaves as regards the future on the basis of existential experiences actually lived through, not on the basis of more or less rationally foretold abstract expectations.[23] For the Algerian peasant, however, the future is the tomb (*Azka d'azqa*). In other words, the future is unconceivable in that it is a void it would be vain to try to grasp. Hence the prohibition of any calculation or scientific attempt at prediction. The future does not belong to human beings. It would be unpardonably presumptuous to mortgage it.

Bourdieu's analysis undoubtedly bears the marks of and is linked to the attempts, especially numerous in recent years, to develop areas known as "depressed." It concludes, in fact, rather disconsolately, that planning does not create effective incentives for the local population precisely because it is based on an abstract idea of the future whereas the Algerian moves only in view of tangible results, immediately obtainable and practical. Not only this, but the traditionalist mentality requires that the plan be presented by a person worthy of respect, an elder of the village who attracts the respect of all, like the *pataviri* of the Indian village or, generally, the tribal patriarch. Rather than the abstract scientific aspects of the plan, human relations count, the face-to-face relations on the community level. Nor does the discussion of common convenience carry much weight, as he who has nothing cannot think of sacrificing himself to increase that nothing. Hence derives the final recommendation, rather surprising, to break down the traditional mentality not little by little in small, short-term operations, but rather with an unexpected radical transformation. Unfortunately, one does not see how it could avoid an essentially, and blindly, technocratic result.

Perhaps an in-depth examination of the religious background to which the Algerian population is linked would be helpful. If we compare the conception of action prevailing in the Protestant ethic with that of the Muslim religion the gap appears obvious. The Protestant ethic, especially in its Calvinist version, is as deterministic as the Islamic. But the practical corollaries are antithetical. Max Weber makes this clear:

Not idleness and enjoyment but only action serves, according to the will of God made manifest, to the increase of his glory. Loss of time is the first and in principle the most serious of all faults. The span of life is very short and infinitely precious to affirm one's own vocation. The loss of time in society, "idle conversation," even the luxury of sleeping more than necessary to health... is from a moral point of view absolutely reprehensible.[24]

To Calvinist hyperactivism and voluntarism, the only relatively sure basis of the certainty of salvation which can never be taken for granted, is counterposed the inertia justified by the Muslim religion, based from the first centuries of the Egira, on a rigidly dogmatic and determinist theological construct, where divine omnipotence necessarily involves the onus of an absolute fatalistic determinism. Sects like the Cadrids and Mutasilids, in favor of free will, as with the development of an Islamic thought relatively heterodox as under the caliphate of the Abassids when Avicenna and Averroes attempted a rationalist interpretation of Islamic doctrine, did not manage to scratch its rigorously orthodox orientation. Islam ended up identifying itself with the dogmatic extremism of the Motakallimum, upholders of the attribution to divinity of a personal, unrestrained power. It is Allah who decides who we are and what we do. Human action loses every importance as independent initiative. One will never plan one's own future nor will its productivity ever be valued. On the contrary, it is believed just and praiseworthy to encourage conversations without an aim or an end, idle by definition and so a waste of time.

For an obvious example, it is considered impolite not to deal and haggle over the price of goods at the time of purchase, not only in the *medina* of little villages but also in the cities and among the richest merchants. Compare this behavior with that in the big supermarkets of industrialized societies, where automatization, the fixed price for every article on sale, the cash register—all is directed at the impersonal rapidity of the transaction so that this can take place in total silence. The contrast appears a radical one. At the same time, the counterposition between the traditional church time and merchants' time[25] worked out by Le Goff collapses. It is scarcely necessary to note that the devaluation of time looms as a fearful obstacle on the road of modernization and, still more precisely, on that of industrialization for those countries of ancient civilization in which the practical application of theoretical knowledge has never been much cultivated, and indeed has been mostly considered as demeaning.

Still today it can be argued that the process of accelerated industrialization in Iran, for example, encounters social, religious, and cultural obstacles that are in fact insurmountable. In premodern societies the conception of time appears as a concern exclusively centered on the present. Tomorrow remains the inscrutable domain of the gods. It is, on the other hand, difficult for peoples who live by hunting and fishing or still as nomadic tribes, to work out programs for the future. Chance, that is, natural phenomena, favorable or adverse, is the force which dominates them and prevents them from having any possibility of prediction.

SYNCHRONIZATION IN THE CLASSICAL WORLD AND IN INDUSTRIALIZED SOCIETIES

Ancient Rome did not restrict herself to exploiting the colonies, but incorporated them, provided them with a civil administration distinct from military justice. She provided them with basic public services, such as aqueducts and roads, and had a relative respect for religious autonomy. Contrary to modern imperialisms—all in varying degrees infected with apartheid—Rome guaranteed the rights of citizenship to the point that Saul of Tarsus could declare, when the need arrived, "*Civis romanus sum.*" Neither piratical coup nor purely technological exploitation, Romanness grasped the basic functional need to synchronize the times of the empire, as I noted earlier regarding the regulation of time as basic tool of power.[26]

This synchronization should not, however, be mistaken for or confused with the modern rationalization of time, an essential characteristic of industrialized societies. Synchronization in the classical world is resolved into the simple statement and practical application of a common regulatory principle, which thereby makes possible the juridically consistent administration of vast masses of humans, heterogeneous by their very nature. It is an administrative instrument, like taxes, military service, or the census. Contemporary synchronization is something different. The synchronization of the Roman Empire is not comparable to that of modern factories and offices. Today, synchronization has entered into the folds of work and the supply of individual energy. It controls times and movements. It constrains to a relatively uniform rhythm, determined and controlled from the center. One is no longer just taken and tied to a general framework in the globality of a society dominated by the same calendar and the same time zone. The determination of times has involved, at least since F. W. Taylor and after,[27] working activity in all its phases, with clear repercussions for life outside the enterprise. One works on the basis of a given time. One lives on a given time. One moves within ever tighter times. The anguish of deadlines spreads together with the cult of precision and speed. Modern synchronization, in its increasingly coercive aspects, leads us to consider the course of life, in that today there is a risk, rather than of living, of "being lived."

NOTES

1. O. Weininger, *Geschlecht und Charakter* (Wien/Leipzig, 1923), p. 152. I borrow the quotation from the introduction by G. Gargani to L. Wittgenstein, *Diari segreti* Rome–Bari: Laterza, 1987). Cf. E. F. Amiel, *Frammenti di un giornale intimo* (Turin: UTET, 1931), p. 36: "Our life is a gestation. The corn we sow by dying is our soul. . . . Thus we forge our destiny."

2. A. Gargani, introduction to *Diari segreti*, pp. 20–21.

3. Augustine, *De vera religione* 39.72.

4. Augustine, *Le confessioni*, chap. 10, emphasis added.

5. Cf. my *Teologia per atei* (Rome–Bari: Laterza, 1983), chap. 2. However, the specific treatment of this is to be found, in psychological terms, in Erik H. Erikson, *Young Man Luther* (New York: Norton, 1958), esp. pp. 203–6: "A revelation, that is, an unexpected burst of internal light, is always linked to a refusal, a cleansing, a rejection; and it would be wholly in harmony with Luther's great freedom on this matter if he were to experience and express it in frankly physical terms. The *cloaca*, at the 'other end' of the corporeal Ego, remained alive for Luther, at times jokingly and at others painfully or depressively, as if it were "dirty ground" where one met the devil, in the

same way one meets God in the *Seelengrund*, where pure being is created."

6. Arthur D. Nock, *Conversion, the Old and the New in Religion from Alexander the Great to Augustine of Hippo* (New York: OUP, 1961).

7. L. Bottani, "Tempo, 'Erlebnis' e significato," *Filosofia* 28/2 (May–Aug. 1986): 147–148. In my view the crucial experience is the transforming one. One is mistaken to believe that simply idling the authentic can convert the inauthentic into itself. That does not take account of the weight of the context and the fact that the dialogue does not occur in a social void, but rather in a commodifying society where *Erlebnis* is experience relived in memory whereas *Erfahrung* is experience as "encounter".

8. M. Halbwachs, *Les cadres sociaux de la mémoire* (Paris–The Hague: Mouton, 1975 [1925]).

9. W. Benjamin, *Erfahrung und Armut*, in *Gesammelte Schriften* (Frankfurt am M., 1972) 2, 1, p. 219.

10. R. Bodei, "Le malatte della tradizione," in F. Desideri, ed., *Walter Benjamin. Il tempo e le forme* (Rome: Editori Riuniti, 1983): 228.

11. Horkheimer's article, originally written in 1934, was published by *L'Homme et la Societé* 69–70 (July–Dec. 1983) as "La metaphysique bergsonienne du temps." It was prefaced by a note by Philippe Soulez which put into perspective the significance of Horkheimer's criticisms and recalling in turn Bergson's merits, especially concerning the historian's work, the philosophy of history changed into prospective teleology, and Bergson's salutary critiques of retrospective illusions.

12. It may seem odd to recall briefly the second part of *Being and Time* where Heidegger elaborates the conception of authentic *Dasein* as *Sein-zum-Tode* (being-for-death), in that death as "mine" and only "mine" comes to assume the role of guarantee of the authenticity and the basic phenomenon of life which finds in anxiety (*Angst*)—not to be confused with *Furcht* [fear]) the critical moment of awakening of self-consciousness from the banal drowsiness of the everyday. It is an amazing convergence of two so-distant worlds of thought, brought in common perhaps unconsciously by the shared romantic German heritage and a Pascaliam and Kierkegaardian pessimism in which the late Horkheimer was also to take on frankly theological tones.

13. Ibid.

14. Ibid.

15. As regards "contextualization", see my *The End of Conversation* (Westport, CT: Greenwood Press, 1988), pp. 83–139.

16. See J. Attali, *Bruits. Essai sur l'économie politique de la musique* (Paris: PUF, 1977). This is an attempt, in some ways disturbing, to present an almost postromantic version of the artist—here, the musician—as unconscious clairvoyant and forerunner of new social formations. The possibility of a political economy of music derives from the identity of functions (to invent an order) which brings together the chaos of noise and the chaos of the burgeoning sociality; music is creative of political order because it is a minor form of sacrifice. In the space of noise, it means symbolically the canalizing of violence and the imaginary, the ritualization of a murder replacing general violence, the statement that a society is possible, sublimating the individual imaginations. On the errors and horrors of forced modernization, see R. Jaulin, *Les chemins du vide* (Paris: Christian-Bourgeois, 1977). A. Schwarz, *Le tiers monde et la modernité de seconde main* (Quebec: F. José Augusto, 1982); in a strictly political vein, see G. Girardi, *Sandinismo, Marxismo, Cristianesimo. La confluenza* (Rome: Borla, 1986.).

17. Andrew Duncan, "Can We Buy Time?" *Times*, 2 Jan. 1987, p. 11.

18. See my *Cinque scenari per il Duemila* (Rome–Bari: Laterza, 1985).

19. G. Martinotti, "Vivere a ritmo di orologio," *Il Sole/24 Or*, 25–26 Aug. 1985.

20. S. Romano, *La lingua e il tempo* (Milan: Scheiwiller, 1983), pp. 20–21.

21. See the recent study by Eviatar Zerubavel, *Hidden Rhythms*, Italian translation (Bologna: Il Mulino, 1985), under the title: *Ritmi nascosti. Orari e calendari nella vita sociale.*

22. P. Bourdieu, "The Attitude of the Algerian Peasant Toward Time," in J. Pitt-Rivers, *Mediterranean Countrymen* (Paris–The Hague: Mouton, 1983), p. 57.

23. Bourdieu, "Attitude of the Algerian Peasant," p. 62.

24. Max Weber, *Die protestantische Ethik und der Geist des Kapitalismus*; I quote from the Italian translation by P. Burresi (Firenze: Sansoni, 1983, [p. 263]).

25. J. Le Goff, *Tempo della Chiesa e tempo del mercante* (Turin: Einaudi, 1977).

26. I am pleased to mention here the contribution of Dr. Arianna Montanari at the seminar on the time of Zakopane organized at Cracow University, 16–20 Oct. 1986. It is hardly necessary to recall P. A. Sorokin's remarks here (in *Social and Cultural Dynamics* [New York, 1937] *passim*).

27. On Taylor, Taylorism, and the scientific organization of labor, see my *Sindacalismo autonomo* (Milan: Comunita, 1950); *Il dilemma dei sindacati americani* (Milan: Comunita, 1954); *La sociologia industriale in America e in Europa* (Turin: Taylor, 1960), now in *Sindacato, Industria, Societa* (Turin: UTET, 1970).

3

The Life Course

THE CULT OF PRECISION

Giovan Battista Vico, that most learned man and rigorous philologist, made a mistake in citing his own birthdate in his *Autobiography*. Clearly, at that time the registrar of births lacked the precision it is noted for today. We also do not know the dates of Rabelais' life. Responsibility for this is clearly not ours alone: If today's scholars cannot establish Rabelais' birthdate, we may well ask ourselves if Rabelais himself knew what it was. Lucien Febvre raises this question is his *Le problème de l'incroyance au xvi*^{me} *siècle, ou la religion de Rabelais*: "in Rabelais' time, did people know their own age, or not?" Febvre concludes that it was not common at that time to know one's own age.[1] A person's age did not seem to depend on the quantitative computation of his years. Perhaps it was rather a question of how one really felt, spiritually and physically.

Everyone was at the age he felt, not necessarily that attributed to him by conventional measurement, linked to a general, and thus abstract, calendar, or from the subjectively binding facts of a document linked to the extrasubjective structure of social power, by definition alien and distant. So, not only was precision in the mathematical, quantitative sense not so important, but the very idea that a living, present individual should have to prove with a supporting document his own identity must have seemed to the average person an absurdity. In fact, when we think of it today, it *is* an absurdity. My great-grandfather, who died in our home in 1940, did not remember the year of his birth with exactitude. However, he had a very clear idea of the season of the year when he was born. In his long evening tales, at once slow and tortuous, suggesting the slow, chance meanderings of a river (the river of time?), my great-grandfather, then a centenarian, remembered being born at the beginning of the summer. It was starting to be hot, and the wheat was tall and yellow, interspersed in the field with the bright red of the poppies; the festival of Corpus Domini was near. His time was the existential experience of hot and cold, a quality in the air, a tactile experience—not an abstract sum.

Perhaps, as some people maintain, this abstract sum in fact allows the individual greater freedom, his own cautious snipping-out of remnants of a private time, to be savored and lived secretly in an otherwise global society. This seems to me highly dubious, but we are not lacking in people who maintain this in all seriousness:

> Today, when the development of the timetable in the West is increasingly based on a philosophy evidently economical as regards time, even our private life and the most minute phenomena of social life are influenced by that orientation. The art of "killing time," of using it in the most efficient and swiftest way, and the negative connotation given to any "gap" of waiting or interval, have emphasized the quantitative vision of time and introduced new practices based on the principle of simultaneous consumption. From the "working lunches" where one discusses business while eating, to the cocktail party, to the dinner party devoted to social contacts; from listening to music whilst driving, to TV at dinner, and so on. However, the growing division of labour and consequent differentiation of the different phases of everyday life in modern society have also favored the separation of the public and the private spheres. In the past, private time was essentially defined as a residual category, no more than a leftover as regards time invested in work. Today, temporal rigidity in our professional commitments, *far from being an alienating phenomenon of regimentation*, provides the individual with greater protection for his private time, and thus acts as a liberating element.[2]

The argument is not new, but its groundedness is nonetheless uncertain. It follows the outline of the famous analysis by Georg Simmel of the mental and sociopsychological conditions of life in the metropolis and the blasé nature of the human type prevalent there. The division of labor and specialization of functions is said to allow the individual personality in the metropolitan environment to offer only one aspect of itself in interpersonal relations, and thus avoid being swept away and, as it were, worn out by the excessive number of contacts. The specialization of professional contacts in this case is supposed to act not as a factor of mass alienation, but as a precious protective barrier for the person.

As always, Simmel's argument is brilliant, but not wholly convincing. In reality, extrasubjective determination of space-time channels weighs on metropolitan man even beyond what Willy Helpbach suspected.[3] I discussed his observations already in the immediate postwar period when the subject of human relations in industry was in fashion and the polemics on free time were raging.[4] Human beings are delicate, unitary mechanisms, formed and dominated by a tendentially unitary logic which cleverly contradicts the mass of juxtaposed, chance, heterogeneous, or incongruous elements. For this basic reason, human beings cannot indifferently or with elegant unconcern pass from socially determined, and to various degrees coercive, "working time" to one which is supposedly free, as though it were a matter of changing one's tie or shirt. Later, this problematic tangle was again decisively attacked:

Technological innovations . . . involve the working process . . . and follow one another with a rhythm and in dimensions never before seen in the history of mankind. This does not mean that work in capitalist industry is without meaning for the workers (this is an aspect which would in any case vary greatly according to the development of the division of labour in capitalist production). But it means there are no longer normatively guaranteed links between the sphere of labour, entrapped in a precise temporal dimension, and the rest of social life, which in turn frees itself from the series of practises induced by tradition. The reflecting face of "public time" introduced by the clock is "private time," which the individual freely disposes of but which remains an objectified time in the sense that it is not wholly involved in the practises acquired through social life.[5]

FREE TIME AND MORTGAGED TIME

What seems obvious but needs to be even more strongly emphasized, is that there is no painless possibility of free time in a society where social time and working time are strictly objectified and measured according to the needs of production and consumption of commodities for the sake of maximizing profits. In other words, the free time of the individual as a private being is not in reality free, but like working time of which it is just the other, symmetrical aspect, necessarily determined. In industrialized societies, the weight of the pendulum, in Marx's famous words,[6] "becomes the precise measure of the relative activity of two workers, as it is of the speed of two engines, so that one can no longer say that an hour from one man is the same as an hour of another, but rather that one man hour is worth another man hour. Time is everything, man is no longer anything; it is all *making time count*—there is no longer a problem of quality, quantity determines everything, hour by hour, day by day."

It is odd that an English sociologist acquainted with Marxism should entertain no doubts when unleashing a critique *en bloc* against historical materialism. Giddens criticizes historical materialism (which is certainly open to criticism, but with much more compelling arguments) on three counts. First, the modes of production of material life in tribal or precapitalist societies are not the principle motor of social change and the same is true as far as class struggle is concerned. Second, historical materialism rests on an ambiguous mixture of evolutionary and ethnocentric ideas as well as on an obsolete philosophical conception of science. Third, Marx erred "when he saw human beings as essentially like tool-making and -using animals, and in using this single, very important criterion to distinguish the 'living beings' who belong to humanity from animals. The social life of man does not start and finish with production."[7]

Regarding the first reason, one might adduce the whole of

the literature on the Asiatic mode of production, Marx's own reflections on the consequences alleged to have followed the building of the railway in India, and the catalogue of the types of relations between oppressors and oppressed with which the *Communist Manifesto* opens. One might respond to the second reason with the argument that the Spencerian and Darwinian ideas which historically have infiltrated historical materialism are also its most complete distortion, in that they "biologize" a process which is and is intended to be wholly historical.[8] One replies to the third reason that Marx consistently supported the idea of the omnilateral or multidimensional man, beyond and against the one-dimensionality of the *homo faber* to which Giddens wants to reduce him, clearly ignorant of the characteristic Marxist dialectical tension whereby every concept is also at the same time a concept-limit (*Grenz-Begriff*). Giddens' vision is all in all simplistic and misleading, the product of wide but perhaps hasty and superficial reading.

The problem of time remains. Once it is objectified in the name of productive goals held to be socially desirable, it lends itself to rational calculation and precise counting—in days, minutes, hours—and falls under the sway of a qualitative control without leaving any way out. This translates its intrinsic value into an ever-faster sequential process, while it presents itself to the individual as an ever-scarcer resource. The devaluation of the everyday is the clumsy justification of this situation, which in reality contains the profound reasons for the crisis and its related loss of meaning, which have been spreading since the advent of the industrial revolution. The etymological meaning of the everyday is invoked to emphasize its meaning of repetitiveness in the common sense, that is, what happens every day, with no aura or originality: the eternal return of the identical, the "every day" of existence. Its basic principle, on principle or by simple unawareness, is forgotten or underplayed, that of reflexive and self-reflexive moment: the absorbed probing beneath the external noises and the internal racket of aimless haste, the stasis which makes ecstasy possible, meditation beyond the deadline, nonmechanistic and nonutilitarian thinking.

The idea of Josef Pieper, that leisure is the basis of culture, takes on full meaning in this context.[9] Decades before, Nietzsche wrote that anyone who does not have at least two hours a day at his complete disposal is a slave. Time lacking is always mortgaged time, time removed from quiet contemplation and tranquil reflection. Paradoxically, one could say that missing time is time which denies time. In a Bergsonian key, it has been remarked that

> Time as experience is creativity, and creativity has always been the cancelling of the perception of time. Creative tension cancels the sense of time whereas acquiescence and indifference multiply it. This is true as much for the individual as at the collective level. In fact, rich cultures, those under tension, create new forms of time. Poor, declining cultures suffer time as a limit, as a bond and as control. It is not by chance that the problem of time and above all of the differentiation of its meanings has been accentuated after the crisis of the '70s. Lived time, creative time in experience, allowed the unification of biological, psychological and social time. The impoverishment of experience, on the other hand, has made the break between individual and social time increasingly perceptible, as well as demonstrating the inability of natural time to contain the expansion and contradictions of social time.[10]

Bergson had already postulated a qualitative time outside the normal, quantitatively measurable, temporal coordinates in the framework of general creative evolution. But creativity does not arise from nothing; it needs, in fact, time, to elaborate itself and mature, at the risk of yielding to the romantic myths of strokes of genius which are, moreover, never too far from strokes of disordered madness.

Creativity needs, and grows in, nonmortgaged time. Reference to the 1970s confirms this. At least part of the creativity of 1968 was spurious. It abolished time because it was unable to plan in time. It wanted everything at once, because it lacked the intellectual clarity necessary to establish a hierarchical order, a list of priorities of the objectives it wanted. It felt it was historically right, but it did not manage to understand fully the reasons for its own rightness. Unfortunately, it could not escape the depressing destiny of a great, missed opportunity.[11]

The mere mechanical, dedialecticized counterposing of coercive social time and personal, private time, one's own and therefore authentic time, indicates an insurmountable impasse and leads reflection up a blind alley. Nonmortgaged time grows in the anxiety, consciously assumed and resolved by performing the task, of time abstracted, of time as scarce resource. It is in shortage of time that the subject discovers and realizes his limits, but in it he also recovers his own genuine potential. This is not a certain result. It is the subject's chance, the gamble of his life. The context weighs heavily, but in the narrow margins left open he has the—never assured, always problematic, and thus dramatic—possibility of reacting. The possibility of reaction stands in direct relation to the capacity for choice. In this sense, the weight of the context expressed in collective times and the blind deadlines of bureaucratic-institutional rhythms do not necessarily involve the process of desubjectivation. This process and its danger exist in mass society, dominated by the great formal structures where rationality has abandoned the individual to make itself the abstract characteristic of impersonal organizations, but they are neither fatal nor irreversible. It is an open question.

The rigid counterposing of social time, mortgaged by collective demands and thereby inauthentic, against time as not quantitatively calculated nor precisely measured, has led philosophical and sociological analysis to an impasse as desperate as it is meaningless. Here, there is no help from speculative, metahistorical, and abstract oracles. They can only supply the consoling help of a kind of philosophical last rites. And yet even the more consistent Cassandras have quickly become aware that man as historical animal lives and realizes himself in time, is nothing outside time: "Out of time—a phrase without meaning—existence can have no sense. It can be neither experienced nor thought in a meaningful manner."[12] The reason may be quickly guessed. If the first expression of the meaningfulness of the experience of the living is to be concerned with and about something, it is obvious that this concern must use time, cannot but realize and develop itself in time, within a "time horizon"—a phrase I prefer to translate into the analogous if not exactly equivalent "historical horizon." One then comprehends the conclusion by Heidegger: "Temporality is the primordial meaning of the being of *Being* (*Dasein*)."[13]

TIME AS A SCARCE COMMODITY

Philosophical weakness, linked to a certain degree of unawareness, has prevented social analysts from using Heidegger's profound intuitions without submitting to their paralyzing fascination. His reflections on memory are certainly valuable. They help us understand how without memory there is no culture and, on the other hand, how without culture or a complex of experiences and shared and lived values, society is not possible. In this sense, tradition is essential. The roots of the past are a guarantee of meaningfulness for the future. The statements and behavior of the 1968's revolt reveal in this connection their characteristic mindless superficiality. In this they were like the vociferous antipastism of the futurists at the beginning of the century, who proposed to destroy the museums while exalting speed, the slap, and the knife-thrust. Not only is the venerating attitude of the student of antiquarian history about whom Nietzche wrote at stake. The question goes deeper. It involves the awareness of one's own original mold, the promises already contained in it and expressed in a potential state. Devotion toward the past in this sense has nothing passive or blindly subordinate or merely repetitive or dogmatically obscurantist in it. Memory is an attitude of internal disposition toward the careful consideration of what once was and what is today. In this sense, memory is an attitude of devotion—one of respect for everything that by giving meaning to life should be preserved in a safe place, away from the hubbub of the market and its transience.[14]

Societies unilaterally oriented and spasmodically straining toward the future are not only destined to lose their past. It is probable they will also lose their future. Already they run the risk of the insignificance of a mindless, arbitrary present where the capacity for evaluation is reduced and the new comes to coincide with the different. On the other hand, the different resolves itself invariably into the already-known. Therefore, for example, a country like the United States is continually changing; it moves from one revolution to another, always prey to profound transformations, whereas in reality the change does not exist, does not go beyond the morphological rind of situations. Structural and sociopsychological changes do not take place.There is only transition from the same to the same. Everything is co-present, simultaneous and flattened. The perspective, sense, and perception of depth are lacking. Analysis splinters against a facade of close-ups. Reasoning is lost in the uniform forest of images. Factuality replaces critical reflection. The terms of questions cloud over. I have already remarked on a singular fact of North American life. With mass television a characteristic process of derealization has begun. What seems in the shopping mall a magic moment outside any determinateness of time and place tends through television to become a mass experience, a collective sleepwalking.[15] Gore Vidal's remark may seem unduly generalized and nasty ("America is a Ramada Inn from coast to coast"), but anyone who has traveled for long, preferably by car, through the endless plains and central deserts of the United States has known first-hand the sense of worn-down insignificance, formless and casual like a landscape after a bombardment, of space void of humanity. They will have known the wide streets lined with the ruins of a pulverized myriad of irregular, fragile buildings, similar as a whole to the painted canvas flats of a touring theater to be struck the day after, in certain urban landscapes, together with the sinister, funereal peace of certain suburbs, unreal in the manicured perfection of the lawns. This is the world of the inauthentic, the world which prefers the comfort of the disposable to that of duration through time, the convenience of superficial change to the pleasure of the stability wherein experiences and meanings sediment.

Escape will necessitate suffering. It is the great historical tragedies, the anguish of failure, the subtle, entangled uncertainties in the face of the difficulty of the task which lay bare the nerves of self-consciousness and thus of the authentic. So too it is the sense and the acute awareness of one's own finiteness that, along with the sense of limit and of one's own mortality, give the individual the awareness of his own irreducible unrepeatability, his own unfakeability, of his final, noninstrumental nature. But the transition from the person as simulacrum to the personality as specific to the person is impossible by way of the pure, labyrinthine introspection which has been bestowed by the philosophical tradition. The initiative whereby the personality of the individual is expressed is not to be characterized through a generic commitment or engagement. To be specifically operational and not be reduced to the outdated alibi which confirms a practical inertia, initiative must come to terms with the historically determinate characteristics of the extrasubjective context. These cannot be deductively foreseen. They must be examined, described, explained, and interpreted by conceptually oriented empirical research. The traditional philosophical outlook with its apparatus of ahistorical concepts and essentialist quiddity here meets its limit. The principles of personal preference and subjective idiosyncracies tend to mask each other and take on the character of universal, and by nature unattainable and untouchable, principles. The specific characteristics of the context become ungraspable. The suggestiveness of the philosophical formula then conceals a cognitive void which blocks self-expressive operational initiative insofar as it ignores the effective historical terms with which the latter must compare itself.[16]

In industrialized societies, possibly for the first time from the historical viewpoint, time becomes a scarce commodity, calculable quantitatively beyond its intrinsic quality. The cult of speed as a good in itself develops, that is, as a goal worthy of being pursued as such. The rhythms of the machine tend to replace human ones. Deadlines reproduce themselves. Temporal pressure generates itself. A generalized hurry makes its entrance in human society, anxiety-producing and frantic, whose reasons and ends have been forgotten.

Seemingly modest authors, in reality detectors of time like sensitive seismographs, are still instructive on the threshold of the year 2000.[17] This is the case with Mario Morasso, essayist and writer from Turin, active in the last quarter of the nineteenth century up to the First World War. His works range from the properly sociocultural, like *Contro quelli che non hanno e che non sanno* (Palermo, 1899), *L'evoluzione del diritto*

(Turin, 1893), *L'origine della razze europee* (Turin, 1895), *Uomini e idee del domani—l'Egoarchia* (Turin, 1898), to subjects still today not lacking in relevance, like *L'imperialismo nel secolo XX* (Milan, 1905), *Il nuovo aspetto meccanico del mondo* (Milan, 1907), *La vita moderna nell'arte* (Turin, 1904), and *La nuova guerra* (Milan, 1974). Morasso was a feverish forerunner of social developments to come (emblematically, one of his poems is entitled *Profezia*), and precisely for this his basic characteristic is at the same time that of spy and bearer of the time he lives in. This is the Italy of the last years of the century, after the tremendous economic and sociopolitical crisis of 1900, which starts decisively along the path of industrial development during the Giolittian decade. It opens up to the great myth of progress as the necessary and inevitable result of technicomechanical evolution. Morasso noted:

> The metropolis, the central fire of civilization, the tempestuous seat of modern life, the crazy terrain of the crowd where the struggle unfolds with a terribly compounded rhythm, the metropolis of endless streets and endless noise, where passes, unstoppable, the tempestuous, inestimable parade of human activity and wealth, the most enormous construct of our age; the amazing pile which only our growing energies have let us build and let us live in—the metropolis is the lighthouse of civilization, the goal of all our desires and all our efforts.[18]

This is a dithyramb to the industrial city, center of the production of technique, which foretells and precedes Marinetti's fiery writings. The city, along with the automobile, arises as a symbol of modern life. It appears as the trademark of a whole civilization, mirrored in it and recognizing itself therein. Humanistic contemplation and the archaic, rustic nature of the traditional culture are replaced by the powerful myth of an urban world, no longer artisan and idyllically rural, but technical, swift, tense. It is no longer tied to the rhythms of nature, but ready to transform it, to invent a new form of life. There is nothing in this regard that is extemporized in Morasso. His will to power, which has often led scholars to mistake him for one of the many fin-de-siècle nationalists or one of Nietzsche's proto-d'Annunzian enthusiasts, takes on an original conceptual form in his works. In the preface to *Uomini e idee del domani—l'Egoarchia* (1898), he says:

This is not a matter of Nietzsche, his superman and his system of moral reversion. Only the customary dilettante will be able to bring forward the absurd accusation of the superman—I give a shrug, and don't bother about them. I admire Nietzsche, and admired him when no one in Italy yet knew who he was, but I firmly assert that my insights on universal aggregations, and on the sexual origin and the evolution of human society have absolutely nothing to do with Nietzsche, and in any case they much preceded the knowledge I had of him from his writings. These ideas, which have an exclusively scientific and positive basis, are mine. Only mine: and finally even my individualistic theory, if it did reconnect with something in precedence, can be found to have its connection with Stirner. My theory is the further development of philosophical individualism.[19]

Possibly this candid statement of faith would suffice to make the limits of Morasso's thought comprehensible. This thought does not transcend the framework of the culture prevailing in his day. It is essentially positivist, evolutionist, basically Spencerian, hardly aware of the historical character of human phenomena. However, what characterizes Morasso's thought and sets it apart as regards Spencer's many Italian epigones, and those of his evolutionist organicism in Italy at the end of the last century, is his ability to grasp the aspects of novelty connected to the new productive techniques. More precisely, his is a real gift of understanding and perceiving, often intuitive and artistic, if not quite explaining and calculating, the social effect of technique, his foreseeing—even if still in an imaginative rather than scientifically specific way—its weight in daily life and on the masses' mode of thinking. Perhaps in this area one can maintain that Morasso remains original also as regards the futurists, who provided him with subjects and metaphors. In this sense, Morasso has a journalist's nose and the eye of the social investigator. He lacks the discipline of scientific research. He is a kind of sociologist without method, incapable of disciplined subjectivity, but his sense of the social problems being created and growing from the apparently chaotic bosom of current life is undoubted. The work that most convincingly demonstrates these gifts, even though with undeniable discontinuities, is certainly *La nuova arma (la macchina)* (1905). It has been remarked that

a true aesthetic of the automobile was asserting itself in France, and also in Italy, especially in the work of Morasso, author of *I prodigi* (1898), *Profezia* (1902) and *La nuova arma* (1905). In this last book, Morasso wrote with precognition of the coming of "new, very powerful beings, of an unknown species, centaurs of flesh and iron, wheels and limbs." The automobile, which should be for youth "a school of will and energy" restores a sense of heroism to the modern age. At the wheel of his automobile "the individual finds his energies multiplied a hundredfold by the machine." Finally the author lyrically described a race in an automobile. It was the same subject dealt with by Marinetti.[20]

The remark is certainly well based, but it repeats and echoes a widely shared opinion. It does not exhaust the question. Morasso's originality is to be seen in his ability to connect the coming of mechanization on a vast scale with the life style and mode of thought, according to a universality of effects and series of repercussions which cross over the level of literature and art to involve the world of social relations and economic and political structures. From the first part of *La nuova arma*, dealing with the rhythm of life, to the sixth part devoted to the man of speed and the heroes of the car, Morasso is never content with merely literary or aesthetic considerations, but on the contrary penetrates the everyday and asks himself about the interconnection between the growing mechanization and quality of social life. He asks about the human types which will be required by the growing number of cars, about the new psychology the car will make necessary to enable adaptation to its rhythms, its operations as impetuous and irresistible as they are impersonal. In its earlier phases, Morasso's thought seems aware of the problems the general use of the automobile creates in the relations between social groups and persons:

What is the maximum purpose of those participating in a sporting event held on a track? To race as fast as possible, to concentrate all their energies and faculties in speeding up the race so as to arrive first at the finish. And they race. race and turn as though they had to escape from some terrible danger, to achieve the single end of arriving as fast as possible back where they started from. Well then, this is the picture the explication of human activity offers us. In the majority of his actions, modern man, even in those related to the intimacy of his affections, seems to have no other aim save to do them quickly. . . .[21]

This critical emphasis on haste without a goal that seems to characterize modern life is of extraordinary significance: it is not an essay in literary bravura, but a sociological observation.

It is this particular sensitivity of Morasso which should be explored. We confine ourselves here to some preliminary remarks. More than once in his works, Morasso shows he knows how to evaluate the impact of the automobile in its various aspects on social life. In the preface to Jean du Taillis's *Da Pechino a Parigi in automobile* (1907), Morasso writes that "what the person feels riding in a 40 HP differs notably from what someone in the 10 HP feels. The sensation of effort is different and the road surface and even the countryside look different. The outline of life changes radically. Hills have a different importance: judgement of distance, because of the difference in speed, changes likewise. The dimensions of the country change. Not only does one seem to see in a different way, but things and events are no longer the same."[22] But Morasso does not stop at mechanical or physiological dimensions. He tries to establish the sociopsychological effects of the automobile:

Here there was another factor of difference: the *state of mind* and thus the sensitive and perceptual activity of Borghese on the one hand, and those following him on the other was not only dissimilar but almost in opposition. The former had every faith in the triumphant efficiency of his equipment and effort; going on as he had begun, he was now certain of victory. The latter, more than being concerned about arriving were downright obsessed by this unattainable Borghese who preceded them. . . . With this book, therefore, the Italian reader will have the complete story of the memorable crossing. He will know how the two little 10 HP De Dions driven by Cormier and Collignon and the 15 HP Spyker kept going in the groove cut by the

40 HP Itala of Prince Borghese, and will be able to draw their profitable lessons—not those declaimed too hastily the day after the victory on the proof of the omnipotence provided by the automobile, and on the brotherhood of peoples, and the car propaganda produced by the run. Nothing of that. . . . (The Peking-Paris) gave not for the automobile but for man the demonstration that he passes over everything and overcomes every obstacle, be he Italian or French, be he at the wheel of a magnificent 40 HP capable of going at 80 km an hour, or a modest 10 HP . . . because he wishes it tenaciously, firmly, irremovably.[22]

The machine, especially the automobile, has a function which is essentially pedagogic. It teaches one to live virilely, like men. At this point it is easy to introduce the transition and understand how war too, in modern conditions, so in need of technical and mechanical devices, can become not only the sole "hygiene of the world" as Marinetti would say, but also the occasion for the development of certain virtues that Victorian morality believed basic: courage, the spirit of solidarity, the sense of renunciation and sacrifice.

In Morasso's foresight, the automobile was destined to supplant the horse. The real race would no longer be between horses but cars. The horse was destined to become an anomaly. The car trip would be seen as a great individual freedom, while that by train is necessarily collective, tied to fixed timetables and predetermined itineraries, and would thus be lived as slavery. Here it is hard to deny the basic justice of Morasso's previsions, the more so as he foresaw that for commerce too the truck would be preferable to the railway. In a word, Morasso correctly foresaw how the automobile would of necessity transform our life.[24]

THE TIME OF THE INDIVIDUAL AND HISTORY IN T. S. ELIOT

The life course develops within a context. It bumps against it. It carves its singularity within the coordinates of the age in which the individual finds himself living. Between the individual life course and the characteristics of the historical context there is a constant confrontation. This is both dialogue and clash, victorious attempts at an interpretation and the triggering of the signals of time—and the bitter recognition of a defeat which from the beginning has appeared inevitable. The time of the individual as personal destiny, and the time of history as the context which bridges the biographical span of the individual, unwind in each other and confront each other in a dialectical-relational process, which is essentially problematic and a priori cannot be profiled. Perhaps only poets can at times decisively articulate that awareness of the problem which today's social analysts do not seem to give thought to. For an obvious and especially instructive example, we shall look at T. S. Eliot's *Gerontion.*[25]

In Eliot's opus this poem occupies a crucial position. It follows the early poems of the *Prufrock* period and precedes *The Waste Land*. *Gerontion* is a basic moment in the development of Eliot's poetry. As such it can only be understood when situated between the early works, *Prufrock*, and *The Waste Land*. Removed from this context, *Gerontion* would seem obscure, would break up into isolated images, losing any internal logic. This lack of a guiding thread could be attributed by the reader to the confused mind of the old man who is its protagonist.[26] However, it is clear that *Gerontion*'s importance is not merely chronological, nor is it simply linked to the fact it is a key transition in Eliot's work. Its importance derives from basic internal reasons, that is, from the meanings it expresses and tries to communicate to the reader.

Stylistically, Eliot achieves and consolidates with *Gerontion* his symbolism, his idea that the poet must present himself, rather than as a specific, individual personality as a means of expression of the values and problems of a civilization. But *Gerontion*'s extraordinary importance goes beyond stylistic form. It is to be found in the essence of its message. This message moves characteristically on two levels: a subjective, existential one and an extrasubjective, conceptual one. For this reason it can at the same time involve the problem of aging and of the meaning of life (that is, of an individual life) and also the problem of history and its meaning (that is, the problem of time). The essentially poetic moment is provided by the pathos expressed in the union of these two levels whereby a problem which appears as an immediately individual problem arrives, on the other hand, at the cosmic level. It becomes a problem of civilization.

Gerontion prepares for Eliot's greater work, *The Waste Land*, since through individual crisis we have a kind of manifestation of the crisis of technically advanced civilization: the personification, in terms of an individual destiny, of a crisis which is, on the contrary, general, and from which one cannot escape.

Technically advanced civilization appears in Eliot as a land without life, where men are hollow strawmen. Why? In essence, beyond metaphors and Eliot's allusions, the reply is simple: because they are men without meaning and without goals which transcend them and so unite them in a communitarian bond. The society dominated by technique is a rich one, but tends to lose problematic awareness and thus its own orientation. It is a society without a general aim. It has only internal technical ends. *Gerontion* thus anticipates the basic themes of Eliot's poetry: the crisis of profound values; the crisis of industrial, bourgeois society which is losing the sense of its historical dimensions; the need to recover collective norms, which Eliot sees as possible only in a Christian conception of life. Eliot seems to propose the acceptance of a phase of interregnum, suffering, and scorched earth—a kind of Ash Wednesday which prepares for the coming of the new society, as he tries to show in *The Idea of a Christian Society* and *Notes toward a Definition of Culture*.

This complex of problems is experienced in *Gerontion* as a personal destiny. Before our eyes, we see an old man alone in a dry season, waiting for the drought to end, while he distractedly listens to someone reading aloud to him. The details are meticulous and cruel. Stephen Spender was right. Gerontion cannot be a mythical hero. It is true he speaks of ancient battles and famous cities, but he also tells us he lives in a "decayed house," that "the jew squats on the window sill, the owner."[27] This is not all: Gerontion is an idle, inert old man, and knows it ("I was neither at the hot gates"). It has been correctly observed that "this is the consciousness of inertia, of an unheroic past, of a lost myth. Gerontion is the first of the 'hollow men' because he has not acted, and has not opened up to the drama . . . has not been regenerated in the 'warm rain.' "[28]

Not only is Gerontion a lazy, inert old man passively waiting for rain, he has also lost his faith. Eliot makes us feel this loss, making much use of Joyce's stream of consciousness.[29] This technique was to be perfected in *The Waste Land* but already here use is often made of it, so much so that it could be used to describe the whole poem from the opening "Here I am, an old man in a dry month" to the final "Thoughts of a dry brain in a dry season." But the application of this technique is not restricted to the interior monologue; it does not simply pass in review a series of states of mind and more or less loosely connected associations. Here, the stream of consciousness also issues from Gerontion's interiority and is manifested in elaborate, historically evocative allusions, if at times only partly expressed or even implicit. For example, the hot gates at which Gerontion says he did not fight, may be those that witnessed the sacrifice of the Spartan warriors at Thermopylae, which literally means "gates of heat." Through this technique facts of historically different and distant epochs become copresent and connected, and the whole poem is given a perspective which goes far beyond the examination of the conscience of a single old man.

Evoked by the knowing, at times cryptic, allusions, the different historical scenarios emerge, through lightning appearances; they are composed into a picture in motion, exceptionally evocative, and are the central part of the poem. This part is an illustration of the decadence, the real depravity of the springlike May. Common sense is inverted here. May and spring are generally considered to be the time of vitality and resurrection. For Eliot, in the foreground are the sensual appetites and their awakening, the fruits to be divided, eaten, and drunk "among whispers." By whom? Who shall enjoy them? Here Eliot abbreviates cultural and historical distances and in a single powerful fresco or single gallery unites Silvero, Hakagawa, Madame de Tornquist, and Fräulein von Kulp; from the classical hot gates of the opening to Vienna, capital of Mittel Europa.

Spender is probably right when he remarks that

> in this poem Eliot obtains excellent results in an effect where he had always shown great intelligence: in condemning attitudes by attrib-

uting to them names and gestures which are themselves negative. Simply by inventing a name like Fräulein von Kulp he is capable of evoking within us the evil hatred we have felt in a boarding-house—in Vienna, for example—for an unknown person who keeps us awake all night, in the next room, by spitting in a particularly disgusting manner which for us involuntarily expresses their Central European personality.[30]

It has been rightly remarked that the allusions in this poem are mostly drawn from the Elizabethans and that Eliot tries to make the past appear like the present. The past can be present because the memory of it exists. At the same time, Eliot intends to use the past to express the contemporaneousness of the present through contrast.[31]

Eliot's purpose, however, seems to me to go deeper. The man who came from Mississippi to Europe to London to find a stable homeland and put down roots there, was not only seeking aesthetic results, but the solution to profound ethical problems. His galleries of characters, historical or linked to the contemporary, running from classical antiquity to the modern and contemporary age, passing easily from one civilization to another, one cultural and religious tradition to another, stand for the poverty of cosmopolitanism. Eliot, like all conservatives, is seeking his own roots and considers the typical indifference of cosmopolitanism for tradition as the primary source of a decadence which is at once religious, moral, and artistic. It is from the indignant denunciation of the blasé indifference of the cosmopolitan spirit that one can work back to the meaning of the "theological" lines which immediately precede the ironic sketches of Silvero, Madame de Tornquist, and Fräulein von Kulp. Theology and religious teaching in general refer one back to tradition, strengthening ancient ties; they bind individuals to the natural community in the specific sense of territory and homeland. They give meaning to those former rites to which Gerontius's tired thoughts now turn, and which one day were brought to the world by "Christ the tiger" and now are in crisis because of the general historical decadence of civilization. The theological lines are certainly obscure but as a whole they transmit this message. They are lines "swaddled with darkness." In a deeper sense than the literal, they are the "word within a word."

"The word within a word." Here Eliot literally cites the phrase of Bishop Lancelot Andrewes, spoken in the Sermon on the Nativity of 1618. Eliot devoted an essay full of admiration to this theologian and writer in his *Selected Essays*. With the mystery of the incarnation God gave the Word to the world but paradoxically by a child "unable to speak a word." To Spender the whole passage seems pervaded with bitter irony, on the basis of which Eliot is expressing the contrast between the mystery of the incarnation and the infant Christ and, on the other hand, the entrance of Christ as tiger. It seems to me that Eliot only demonstrates a difference, polemical certainly, but not ironical, between the believing age of rituals observed with devotion, and the juvenescence of the new times in which "came Christ the tiger," almost as though to symbolize that union of beauty and terror which lies at the basis of the poetic and human experience of *The Waste Land*.

I agree with other commentators who see Eliot here as a man not yet religious in a specific sense, not yet bound to a particular church, but who nonetheless feels the need of the tie (*religio* means bond), aspires to the community of believers beyond the "depraved communion" or eucharistic feast, cosmopolitan and blasphemous, and hopes to approach (as in fact he would draw near) to an orthodoxy that could give him certainty and assurance.[32] The sacrament of spring, on the contrary, lies in its spiritual and natural sense in *Gerontion* in being eaten in communion among people who whisper (and here Eliot introduces a further element of depravation). The subject is thus that of a radical perversion of devotion and religious rites, a perversion well in view in Silvero's "caressing hands," whose devotion passes from the Last Supper to Limoges porcelain; in the madame transformed into a medium; and finally Fräulein von Kulp who is her client. Gerontion knows this is a depravation, or at least guesses so. In fact, he is spiritually sterile, useless. The wind motif returns: "Vacant shuttles weave the wind." And finally there is his own definition, almost an epigraph: "An old man in a draughty house."

So the world has not recognized its own possibility of salvation; it has forgotten the rites of tradition, perverted the eucharist, and instead of eternal salvation ends up finding history. Gerontion thereby becomes an emblematic figure. He is a specific individual, with his own problems and life, but at the same time he is a singular universe. In his personal history there is reflected, compressed, contracted as it were, the history of humanity. Eliot is thus playing on two levels. Gerontion's words are concrete and rational. They are the discursive words of prose, not the absolute, metaphysical ones of poetry, to use Bakhtin's formula. There is the thickness of the everyday and its greyness. And yet there is also the mysterious, unpredictable breathing of history. Gerontion is an old man and at the same time a symbol. He grows old as humanity has grown old. He lives in a house which is falling to pieces, just as humanity lives in a polluted and degraded world. He is insecure and disoriented like humanity, which has not recognized the sign of its salvation; he is disoriented. There is thus a doubling of levels in Gerontion which intersect or are co-present, and achieve effects of great evocativeness. For Gerontion's story is presented as the highly concrete and specific, dated, and experienced history of an individual, but is presented also as the history of humanity.

But what is history, then? Certainly not historicism's history, history conceived of as a unilinear development, essentially void of dramas, in which the consequent is automatically explained and justified by the preceding, and for which progress becomes a simple chronology of the inevitable. The critique of this model of ingeniously optimistic and progressive historical development was long worked out by Eliot. Everything that for the historicist, progressive mentality was a simple residual survival, the detritus of a now bygone tradition which little by little would be swept away to make room for the advance of reason and progress is, for Eliot, essential. Its best part, and also that which is most specifically its own may be that where the poor dead people, his forbears, assert "their immortality" with the greatest vigor.[33] The history Eliot speaks of is thus not historicist history. It is the history of every individual who in living relives the history of those preceding. It has been suggested that "dead writers are distant from us because we know much more than they did." Exactly, Eliot observes, they are what we know. Beyond any sense of tradition, Eliot sees no possibility of poetry or moral values or political orientation: "What one must insist on is that the poet must develop or secure awareness of the past and must continue to develop this awareness for the whole of his career. What happens is a continual surrender of himself when he is faced with something of greater value. The progress of an artist is a continual self-sacrifice, a continual extinction of the personality."[34]

The history of humanity and personal aging in Eliot thus come into contact and intertwine and influence one another. Eliot's evocative novelty lies in a sense of poetic time which, however, is not universal and extraindividual but on the contrary is linked to the vicissitudes and sentiments of the individual. The poetic word, from being metaphysical, oracular, and timeless, becomes existential word, tied to the rhythms and natural, concrete processes of aging, or to time as the duration of a single human life, and vice versa. It is this dialectic between the history of humanity and life story that makes Eliot extraordinarily modern and emblematic. We can no longer know with certainty where we come from and where we are going. Progress has stopped being a chronological fatality. Everyone, every moment, performs the gesture that saves or condemns. We are in no man's land, the waste land. Eliot is fully aware of the poverty of historicism, convinced that it is enough to go forward to go well, and is furthermore aware of the complexities of history as a human affair: "History has many cunning passages, contrived corridors," he writes. To him history is neither economically nor ideologically determined. It does not have a logical, coherent pattern, as it is guided by "whispering ambitions" and by "vanities." History thus cannot be outlined, as happens in historicist or paleopositivist thought, as a unilinear flow, with fixed stages, without disturbances, and perfectly understandable, from tradition to reason, custom built on oral communication to the abstract rationality of interchangeable, mechanical, technical apparatuses.[35] In Eliot history returns to being in the first instance a human undertaking, characterized by the passions of man, his intelligence but also his vanity and evil, a sequence of episodes not only of cold rationality but also of plots, struggles, and blood.

A characteristic tension in Eliot appears here—the counter-posing of the decline, the crisis of modern society, and the security, the meaningful rites of the society of the past. Doubt assails even he who proposes to preserve the values of the past. In fact, "I have lost my passion: why should I keep it;" terror and beauty, selflessness and pettiness, tolerance and inquisition seem to be historically interwoven to such a degree that understanding and logical explanation become impossible. Why, then, Eliot asks, preserve past experience "since what is kept must be adulterated?"

This is the problem of the coexistence of the old and the new, past and future. How to recover the values of the past, adapt and enliven them in the present situation in the prospect of the future? In other words, is there meaning to human history? Again, what meaning does individual suffering have? The passage of time, its constant movement—must this inevitably be a waste of time, a kind of dramatic (because unstanchable) bleeding?

Time thus emerges as the central problem of Eliot's poetics and thought. This has been well put by Spender: "He was obsessed by time. The past and the modern co-exist in his poetry like an imaginary present of symbols in conflict, to which are connected values of time or spiritual death."[36] But Spender's judgment regarding the last part of *Gerontion* seems to me too negative to be acceptable. After the strong first part of *Gerontion* the poem loses itself in dark corridors and passages. It is more cryptic than difficult. Eliot does not propose to write a didactic poem on the effects of history and the decline of the Christian faith. The subject is only a pretext which is transformed into a poem. The trouble is that the theme is not so much transformed as rather trapped in the highly metaphorical style of Tourneur's *Revenger's Tragedy*. The poem is strange, bizarre, splendid, but ultimately linked to Eliot's conception of the parallel between modern Europe and the Machiavellian setting of the Italian courts as described by the post-Elizabethan dramatists.[37]

In reality, the last part of *Gerontion* has an undoubted dramatic pathos:

> The universe disintegrates, human names whirl beyond the Bear, a seagull flies against the wind and its feathers fall in the snow while the Gulf claims. In context, the Gulf seems to mean annihilation, and refers us to the Shakespearean epigraph, the warning to be "absolute for death." If this is the sense, death is already here—and in the *Waste Land* the sense will be clear from the start, from the epigraph. But Gerontion does not accept death. In chaos, he sees himself driven by the Trades to a sleepy corner. The nightmare too has been only a moment, one wandering among others, more awful, perhaps, but without resolution. The poem ends as it began, "Tenants of the house/ Thoughts of a dry brain in a dry season."[38]

Eliot's thought in *Gerontion* derives from the clear consciousness that history is not supported by a sovereign design, that it is not a plot or a rationally intelligible project, but rather a congeries of ups and downs, chance, blind encounters and clashes, obscurities, and a slow, imperceptible accumulation of rubbish, without logos and without ends. In this sense Gerontion is a defeated man who prepares, in his passive waiting, for the hollow men of *The Waste Land*. This is not, then, so much a lapse of forms or style as Spender suspects. The fall is crucial. It concerns a specific stage of Eliot's spiritual evolution in his progress toward the rediscovery of religious values and his conversion to Catholic orthodoxy. History, in fact, has no meaning in itself. The age of technical progress has shrunk it into a series of fragmented chronological units, without meaning. It can acquire again its whole meaning only as premise to the eternal, accepting its reentry into a providential design as time out of time, or time that does not contain its own justification, and which instead justifies itself only as a preliminary moment of a metahistorical reality, and as occasion for the restoration of the timeless truths of tradition.

Here, the role of the subject, personal initiative, and poetry as an individual undertaking break down. Tradition has a double value. First, it is a factor of psychological assurance, in that it links the individual to a group of moral and spiritual certainties that belong to the community. Second, it is a survival technique. It contains cognitive values, experiences, and practical methods of life that help human groups to resist crises, to survive historically, and to develop. In this perspective, the ego as center of experiences and as value is only a romantic superstition, perhaps a presumption, certainly a barrier, rather than a help, for the production of great poetry. Eliot seems to believe that everyone, by genuinely living and expressing himself, also expresses his time, lives his epoch in himself, writes not only his own history but also the history of his time. For Eliot, the biography of the individual, his private emotions, does not make great, significant poetry. For him, states of mind are like raw material in its crude state, to be transformed and poured into a finished poetic product: "There are many people who appreciate the expression of sincere emotion in verse, and, furthermore, there is a smaller number of people who can appreciate technical excellence. But there are very few who can appreciate when there is expression of a *significant* emotion, an emotion that has its life in the poem and not in the history of the poet. Artistic emotion is impersonal."[39]

BASIC TIMES

If it is true that every individual is a singular universe, one must also recognize that the life history of an individual is also a universal history in miniature. This justifies reflecting on the course of life as it can be deduced from biographical testimony and autobiographical accounts. Biography and autobiography are essentially a journey in time, without return, characteristically irreversible. Possibly for this reason alone, the term *course of life* seems preferable to that of *life cycle*.[40] We are dealing, however, with a journey at variable speed, marked by nonuniform rhythms, completed and touching on a multiplicity of levels, times, and cross-rhythms.

Postponing examination of this subject to another time and concentrating on temporality, we can distinguish four basic times:

1. Individual, existential biographical time, and that of the primary group. The group functions here almost as prosthesis, if not the

placenta, of the individual. This time has three forms, the life history, life story, and autobiography.

2. Historical time, divided into macrohistorical time (long duration [longue durée], generation, and event); politicoinstitutional time (tradition, revolution, and evolution); economicoconjunctural time (business cycles, interest rates ralated to conjuncture, relations of production, and types of labor organization—individual production, production line, piece work).
3. Ahistorical time, or sacred time, which transcends and negates profane temporality as *festa* time, mythic time of the origin of hierophanic time, revealing the rhythms and secrets of the universe.
4. Natural time (nighttime, daytime, and seasonal time).

Existential individual time and historical time are both unilinear times, whereas ahistorical or sacred time and natural time are cyclical times which always turn back on themselves like an enlarged, eternal present.

The course of life, on the other hand, takes the form of a journey in autobiography. It is the traveler who tells himself his own life adventure. Why? What determines the decision to tell his own life, to write his own autobiography? Why does he travel? What is hidden behind the metaphor of the journey? Usually, one goes on a journey to find something. In fact, one undertakes a journey only so as to return. One returns to the starting point, but changed inside, more aware of oneself, having enjoyed the advantage of leaving home which allows an overview of the world, oneself, and others, which is clearer, distanced, less clouded by the mists of familiarity which smudges the outlines of the everyday to the point of making them invisible. One thus decides on a journey because of an internal dissatisfaction, an imbalance. The driving force which underlies the journey is the painful awareness of the loss of an original balance. At a certain point, the pain is no longer bearable. One leaves to find something that has been lost, a mislaid good, but one does not find the same thing again. One finds another, but does not yet know what it may be. In fact, one goes in search of one's own lost, or weak, or confused identity. But in the course of the journey the identity one is seeking inevitably changes, and at the end of the journey one finds oneself with a new, unheard-of identity. One leaves to seek oneself and ends up finding an other, one who bears no resemblance, an unknown, an intruder.

The autobiography-project, that of the young, outlines what one might call the imaginary autobiography, desired, hypothesized as personal career and the whole of the objectives to attain. The autobiography-summary expresses the need to look over one's shoulder on reaching middle age, on the threshold of mature life ("in the middle of our days"). The autobiography–balance sheet produces a profit and loss balance for a whole life now reaching its extreme limit. No doubt the autobiography in this sense is a pause for reflection, and may be the occasion for an important *prise de conscience* for the individual, an essential premise for any plan of self-analysis and self-formation. The first problem regards the ambiguity of recall. The basic tool of autobiography is memory, on which the first part of this book mainly concentrated. As I said, memory is selective. This is true, but not sufficient. In the autobiography, the past is evoked, summoned from its shadows, is made present again, or is re-presented, presentified. It is not the past as it was. It is the past as it is now, recalled, relived in its re-presentation. What returns from the past, thanks to memory is, as it were, a selected, purified, reconstructed past.

I said earlier that memory is the place of devotion. In fact, the past memory evokes, guards, and venerates is not the past of the past, the past in its empirical factual determinateness as one might deduce it from a police report. The text memory works on is not the whole sedimentation of images, sensations, and perceptions accumulated in the cellars of consciousness. Memory works on a trace, reconstructs the general picture, a gestalt form, from the unsteady outline of a blurred footprint.[41] I said earlier that memory is a faculty that forgets, one which one cannot trust, which selects at random, seemingly, and distorts. But a period of history that lets memory atrophy in favor of the calculator and the tape is one which is self-condemned. Now one must recognize and assert that there is something "magical" (in the sense of the root *mag*) in the power of this faculty. For memory is not restricted to working in the frame of a single time. It breaks the banks. It touches on, involves, and joins the four basic times. It may draw a memory-sensation from the first, the sedimentations of natural time and existential individual biographical time, make it play, graft it on the third and make it interact (historical time), and print it, finally eternalizing it with mystical rage in the ahistorical time of absolute extra-everydayness. I prefer to define the latter as meta-everyday, or second-degree, stressed everydayness, whose temporal determinateness is suspended in favor of a nontemporal eternal present. In this sense, individual memory, so disordered and even chaotic in appearance, draws near to and tends to resemble and act like a collective organ of humanity, when it has to evoke and represent a temporal image in the present. It depersonalizes itself and acquires an emblematic value, which transcends it as purely individual memory. The individual is universalized, while remaining individual. Indeed, his individual nature is emphasized, to the point of again drawing on the basic archetypes of the common humanity of human beings.

The paradoxes of autobiography are as follows: (1) autobiography is a personal account which is, however, bearer of collective meanings; (2) autobiography is a private statement which, however, produces images which translate private memories into public identities; (3) one needs to have reached old age to have material for an autobiography, but autobiography is based as an account on seminal or molding images from early childhood and adolescence; and (4) autobiography continually wavers between the inside and the outside, an interior that is mine alone, and an exterior that is made public and as such inevitably escapes me, as my body escapes me, its growth, evolution, aging while, within me, I am always nineteen.[42]

THE INTERNALIZED EXPERIENCE OF SOCIAL CHANGE

Social change becomes an internalized experience, but it is not thereby psychologically diluted. The course of life as biographical time is reconnected to historical time dialectically. The sociological category of social change, so often divided and impoverished by the one-sided examination of its structural or psychoanthropological aspects, regains its profound unity as change in experienced time. I have grown old with Italy. I have become old while a peasant and artisan country has, in the course of little more than two generations, become urban and industrial. The country of the hundred cities Edmondo De Amicis wrote about with lyrical and bitter nationalism is today the country of the five great metropolitan systems (Turin, Milan, Rome, Naples, Palermo). The garden of Europe sells more machine tools worldwide than Sicilian oranges.

It is inevitable that the construction of a life history should always be a reconstruction. To believe that the life history is essentially known through its empiricotemporal causes as it has really developed, would imply rather seriously falling into ingenious realism, devoid of adequate gnoseological awareness. It has been rightly remarked that the construction of a life history is the way the individual represents those aspects of his past that are relevant for the terms of the future-oriented intentions whereby he directs his current actions. Life histories are not thus a collection of all the events of the course of life of an individual. They are, rather, made up of images reconstructed with the aid of memory. One could also say they are reinvented through the recall of images or traces of the past through the re-creative function of memory.[43] The re-creative function of memory, however, needs to count on a personal identity capable of developing in relative autonomy within a safe habitat, where the individual feels at home, not threatened by that foreign land or external power of global society. Then the dialectical relation between private and public can develop in ceaseless play, which sees these interweave and mutually interact.

Industrialized societies, in their technically advanced capitalistic form, tend however to alter the relation between public and private spheres profoundly, so as to modify the normal process of the phases of life as this was traditionally determined. The generally accepted itinerary of life, with its stages running from infancy to old age, seems to be deeply modified, probably irreversibly, by the logic of advanced industrial society.

This logic is all-pervasive. It has its own typical indirect totalitarianism. It tends not to leave empty spaces. It invades and commodifies all aspects of life. It penetrates the deepest folds of interpersonal relations into the very structure of the individual. The identification of the person is endangered. Privacy is invaded. On the other hand, privacy is the necessary precondition for the formation and assertion of the self, to make the self able to resist requests made by others, especially when these are in contradiction with one another. But in a new environment, continually renewed under the pressure of a productive technology in rapid development and whose requests are not yet well known to the individual, the individual tends not to say everything, not to open up, to observe a behavior and have a bearing that are reserved. This can make privacy a burden, even when it helps the self resist an oversocialized environment. Privacy does not then allow the individual, his self, to express himself completely. It forces him into self-censorship. This in turn inevitably produces a sense of isolation and social estrangement which can even reach the torment of secrecy, or the precise sensation of not being able to present or express oneself as one genuinely is.[44]

The endangered personal identity, no longer able to form itself completely and show itself genuinely, is replaced by the ethic of the corporation, or that of the big firm or technostructure which upholds the social texture. The morality of the organization, intended to maximize profit as the supreme criterion of public and private morality, replaces individual morality coherently codified in terms of values, punishments, and rewards. The unique individual is watered down to the grey and interchangeable organization man. The principle of individuation which is at the basis of personality is blocked; it no longer works. The remarks made above are basically valid, but the logic of industrialized societies is altering its terms. It is true that the self is in the world and the world is in the self:

The structure of individual life is an organizational mode of the self and the world. However, the *self and the world are not separate entities*. They are not like billiard balls which, when they meet, influence their reciprocal directions but not their joint natures. An essential characteristic of human life is the interpretation of the self and the world. Each is within the other. . . . The structure of society is reflected in the self and in the structure of life. The life of every man proves both the wisdom and integration of his society and the conflicts, oppression and destructiveness that mark it.[45]

UNIVERSAL COMMODIFICATION

It would, however, be a grave error of method and substance to place self and world on the same level, and view them as two forces in a state of parity. In technically advanced societies, the pressure of the context and of its standardizing, functional requirements risks weakening, if not paralyzing, the formative process of the individuation of the person. The differentiation between external and internal worlds becomes ephemeral. The individual appears fragile and vulnerable. Capitalist logic, with its emphasis on the impersonal functioning of the market as supreme criterion of social regulation and sole moment of collective aggregation, is also penetrating the institutions of primary socialization, like the family, which until recently have resisted its spread.[46] The capitalist system, which once involved only certain key industries like steel, today has invaded the whole of life, commodifying it.[47] From this point of view, the concept of psychological crisis itself must be revised in depth. So too the various phases, or "crises of the life cycle," to use Erik H. Erikson's terminology, are only to be seen as a convenient schema, not to be taken literally. The individual's crises of identity go beyond the capacity of individual control. The mold of individual neuroses does not necessarily break, nor does it depend essentially on the individual's surroundings. It refers to a sociogenetic causal or conditional structure. The problems of the individual are not an individual fact. The rhythms of the machine are imposed on him beyond and against his individual rhythms. The processes of maturing are accelerated independently of the developmental ability of the individuals. The times are ever shorter.[48]

In the same context, however, disturbing phenomena are recorded. Industrialized societies invent protracted adolescence and youth and prolonged old age. The degree of uncertainty and precariousness increases with them. The map of the phases of life worked out by Erik H. Erikson (infancy–early childhood–the age of play and school–adolescence–youth–adult–adulthood–old age),[49] is acceptable, but with the proviso that this is not a matter of automatic gradations or rigid stages of development. Erikson himself recognizes that in protracted adolescence the perception of time can be subjected to serious disturbances: "In extreme cases of deferred and protracted adolescence, there takes shape an extreme form of disturbance in the experience of time which, in its most acute form, belongs to the psychopathology of everyday adolescence. It lies in a sense of great hurry and at the same time a loss of concern for time as a dimension of living. The young person may feel simultaneously very young, really infantile, and old beyond any possibility of rejuvenation."[50]

For Erikson, a precise sense of ego-identity requires that an individual be capable of self-possession, that is, that he be able to define himself in terms of his whole life span. This means that the conquest of identity runs through the ability of rationally planning one's own future. But according to Herbert Rappaport's research, especially among the young there is a huge group of the undecided, caught in the swirl of indecision and existential doubts, undefined but tormenting. These people see themselves more in terms of the past than in future perspectives, and are defined by him as the "moratorium group," the group of stasis or suspension of decisions.[51] As the context changes, so too do individual attitudes. Whereas taking drugs tends to widen and, as it were, eternalize the present, with the aim of guaranteeing a flight from responsibility, the evidence of Philip Zimbardo documents the development of his perception of time along with the change of context: "As a poor kid in the Bronx I grew up with present-oriented people; my father was unemployed, but my relatives came to visit with bottles of wine and mandolines. . . . [At college] I learnt from my profs to defer immediate pleasure so as to get rewards later on; I was becoming a future-oriented person. . . . Later, however, I got obsessed by the future. I had no time for my wife or my friends. If someone called on us without warning, I considered it more of an imposition than a pleasure."[52]

Only a systematic collection of life and career histories, with special attention to the perception of time according to the various stages and different positions reached, could provide useful empirical material for further reflections on this fascinating subject. Bearing in mind that the developmental capacity of the individual lasts and develops throughout life, it seems possible to put forward a schema simplified into five basic life stages: childhood–adolescence–youth–maturity–old age.

Each of these stages has its typical problems and specific perception of time, to be investigated with empirical research and critical reflection, removed from any romantic exalting of change—bearing in mind that decomposition is also a change. Industrialized societies, as distinct from other kinds of society and culture, make adolescence and old age especially problematic. First, they "create" them, as we have seen, with refined techniques of knowledge, which slow down full-scale entry to the labor market. They tend to raise the level of schooling, and in turn considerably extend life through medical care. Paradoxically, then, at the very moment when the young are ready to take on full responsibilities and the old people see their years of physical and mental capacity lengthen, the former are parked in the limbo of a protracted adolescence, and the latter see themselves pushed aside, since their experience, in view of rapid technical changes, is considered useless, if not damaging. Experience is devalued and considered a disadvantage.

PROTRACTED ADOLESCENCE AND THE CHALLENGE TO INSTITUTIONS

A technically advanced, opulent, and permissive society allows everything but does not suffer the poverty that raises desire. It is a society where adolescents and young adults run the risk of an underdevelopment of the imagination, its disappearance through simple atrophy. Certainly they can recount their few, raw experiences. They do not know how, are unable, or have not the interest, to tell their imaginary autobiography, project themselves into the future, or gamble on destiny. They live by the day or the minute, closed in the present as if in a prison. On the other hand, adolescence is a typical invention of industrialized societies. It expresses, in the terms and limits of personal destinies, their ability to sustain non-productive individuals for a considerable period of time, which, it would appear, is tending to expand. In the European middle classes the children living at home more or less at their parents' expense until they are about thirty are no longer an exception.

In paleotechnical and preindustrial societies the step from childhood to young adulthood, on the other hand, was very short. One went from elementary school, when these existed and functioned, to work in a span of time that is today wholly taken up with school. Work for minors was the rule not the exception. Chapters 12 and 13 of the first volume of *Capital* should be reread not so much to arouse moral revulsion but to recover an objective picture of the situation at the origins of capitalism,[53] at the moment of the transition or "great transformation"[54] as it has been called, from the artisan and peasant world to the technically advanced environment, urbanized and industrial.

In late capitalist or postindustrial societies, as it has become the fashion to call them, adolescents call for full responsibilities, worthy of mature persons, and instead are given useless, costly academic courses, often taught by demotivated teachers, paid a fiftieth of a mediocre footballer's wage in a culture where financial reward and recognized social value tend to coincide. Adolescents are, as it where, in an eternal waiting-room. It should not be unduly surprising that adolescence then appears as an intrinsically hybrid phase of life, in some way unnatural. To fully grasp what I mean one should read Stendhal's *Memoirs of an Egotist*, Thérèse de Lisieux's *Histoire d'une âme*, and Colette's *Naissance du jour*.

To the superficial observer, modern adolescents seem quiet, pragmatic, in need of understanding and affection, very different from teenagers twenty years ago, who wanted to change the world. They know nothing can be changed at once, through a sudden inspiration or unforeseen outburst of protest. They seem calm and resigned. But is it really so? Their distance from the adult world, like all appearances, is perhaps more radical than the noisy rejection and turbulent contestation of 1968. It is said today that psychological dimensions prevail and that the private wins. Perhaps. But it may also be that we have expressions in pscyhologicoindividual terms of structural contradictions, before which today's adolescent quickly sees he can do little. His calm is not resignation before the fact, still less egocentric opportunism starved of personal and professional assertion. It is the conscious strategy of one who knows that, being caught in the grip of an objective structural contradiction, it is better not to get excited. Rather, one should stay totally motionless to avoid further suffering or dramatic worsening of the situation.

The rejection of the adolescent is thus not in itself deviant or irrational. Every generation needs to break with the mental climate, modes of thought, and behavior of the previous generation, if only to assert oneself and guarantee one's autonomy. This process of self-assertion necessarily implies painful transitions and periods of great, tormenting insecurity. The conquest of one's own specific identity is a difficult crossing of the desert, the more so as the process of self-identification can never be said to be concluded. It is a process of development that lasts a whole lifetime, and no one acquires an identity without being exposed to crises, occlusions, more or less serious and tormenting corrections, and revisions. The cases mentioned by Erik Erikson concern famous people, but those of ordinary people are equally complex, intricate, and at least on the surface confused.[55] They can lead to the formation of a negative identity, which may further lead to a situation of paralysis of subjective initiatives tied to a deep sense of inadequacy which does not necessarily reflect a situation of real incapacity.

This sense of existential inadequacy, wherein the sense of time appears contradictorily as an absolute lack of time and as empty time, without commitments and deadlines, may depend on an unreasonable ego, which does not accept everyday reality, the world of common sense and reasonableness, and which wants omnipotence and omniscience at once, here and now, with no effort. In other words, the adolescent, in his search for self-assertion, can fall a relatively easy victim to a perfectionist demanding, and persistent superego. There are not many ways out. One can only give up liberty as the property of oneself, or self-possession. One can try to live without projects, sense of time, precise orders, or on a multiplicity of levels at the same time as though one could in fact live many lives at once. This should not be confused with the multiple richness of overlapping loyalties of a mature person who regulates his choices. Rather, it is the rejection of existential coherence, once the scantiness of available resources has been determined; the psychic, external tendencies, simultaneous and contrasting and therefore tormenting; the impossibility of a meaningful synthesis in the face of the chaos of suggestions and impulses. It is the rejection of meaningfully planning one's own life. It is a renunciation which is not simple or easy, and to which there is probably linked that unmotivated melancholy which so well expresses existential uncertainty, in the same way that the first down of puberty shows the mobile, shifting borders between childhood and adulthood.

To interpret and correctly evaluate the secessionist attitude of today's adolescents as against adult society and its formal institutions,[56] the opinion poll is only partly useful. Its procedure is too direct and mechanical, especially its precoded questions. In some cases it can be intimidatory and unconsciously push interviewees to lie protectively. On the other hand, the collection of life histories is not easy. It presupposes analysis and the critical transcendence of current epistemologies,[57] and at all events it requires a lot of time. My research on youth and drugs (*Giovani e droga*) fully confirms this.[58] And yet, the poll recently sponsored by the Italian Association for Demographic Education (AIED) has obtained results worth pondering.

BETWEEN RESIGNATION, REBELLION, AND UTOPIA

It seems reasonable at the outset to see adolescence as an invention of industrialized societies, societies which have in fact a need to provide their young people with quite extensive, specialized technical knowledge, and which therefore tend to raise the number of years of compulsory schooling. But one of the basic problems of adolescence today is precisely that provided by a strange contradiction. Today the process of maturation seems considerably faster compared with previous historical epochs. Yet at the moment when young people feel ready to take on responsibilities of their own, have a family, and manage their life autonomously, society as a whole forces them to mark time, wait, not plan, continue a process of preparation or primary socialization increasingly seen by those involved as a suffocating one. It is this contradiction which seems to provoke the descriptions and interpretations of adolescence most widely known, from the literary ones like Musil's *Young Törless* or Mann's *Tonio Kröger* to the psychopedagogical or sociological ones. Adolescence is generally described as indecision full of a sense of destiny, a libertarian impulse, a magic moment of the life plan or original vision Joseph A. Schumpeter speaks of. Moreover, we find this in Auguste Comte, where he says his famous law of three stages was essentially intuited by him at the young age of fourteen. In this perspective one understands the radicalness of the adolescent in rebelling against family ties and the institutions of primary socialization. The adolescent wants to destroy everything to assert himself. He is in the phase when all childhood identifications and conditioning are put into question so as to realize the completeness of personal identity according to criteria of authenticity and self-assertion. The original self, formed in infancy, moves toward a situation in which it will be able autonomously to face the outside world, independently as regards the family, the school, received teaching, and traditional values. If, as I believe, things are like this, it seems reasonable that the poll as a research instrument is quite inadequate by itself. The questions, asked by adults in a rigid questionnaire, can indeed appear as a further form of violence, against every good intention. More refined research methods, especially in the first delicate phase of the research, bring out the problems not as they are seen by adult investigators but rather as they are lived by the adolescents themselves. The method of life histories then becomes crucial. Quantitative methods will be called on, with suitable weightings, to check the hypotheses that emerge. Within these limits, and despite these well-founded reservations, the research sponsored by AIED[59] has considerable merit and provides results both of method and substance worth thinking about. In this regard, I shall offer some methodological remarks and then look at the basic questions.

The number of young people interviewed was exceptionally high. This fact should, however, be taken with some caution with reference to the representativeness of the sample. In fact there are some shifts in relation to the census data. For example, the number of adolescents residing in the central region is high (45 percent) compared with a census population of 25 percent. The presence of residents in the North is correspondingly at a disadvantage (30 percent instead of 45 percent); there was a good fit for the group interviewed resident in the South and Islands.

The age distribution, acceptable for fifteen to seventeen year olds, is very heavily weighted toward eighteen to nineteen year olds (45 percent of the total), and inversely for the younger (10 percent).

The male-female distribution overestimates the latter. Furthermore, there is no record of the size of the communes of residence.

These remarks do not condemn the work done, but rather call on the investigators to give qualitative, not purely statistical information on the methods of collecting interviews. On the other hand, it should not be forgotten that in the reading of the following data the replies are from a clear majority of adolescents over fifteen, who live to a large extent in central Italy.

It should also be borne in mind that we are dealing with a good majority of young people who do no paid work and are still involved in academics: 54 to 56 percent are students, but in this case such a high percentage is not surprising.

The focus of the research is on relations in the family of origin and sexual relationships. Far in the background, there are other dimensions (religion, politics, work). In the private dimension, the questions focus on sexual and emotional behavior. One must ask if the data gathered are adequately balanced, if they are readable, in other words, from a sociological point of view. Perhaps the sociocultural context in which these twenty thousand young people live stays too much in the shadows.

Both sexes respond that they belong to families whose economic level is perceived by them as middle-high-middle (80 percent). For 90 percent, their parents are married and living together, and so few have the experience of separations and divorces. Almost all unmarried, these young people live in the family-kin nucleus. It is less clear who the males (210) and females (576) who say they are married or living together live with. One guesses that only a few live alone and that others share their home with friends or relatives and parents.

The family nucleus takes care of providing both sexes with the money they need to satisfy personal needs (85–90 percent).

The study thus confirms what had already appeared in other specific studies on family budgets, where it seems that today the family nucleus tends to be not only a unit of consumption but also a place in which a redistribution of incomes is carried out, with very important results. In fact this behavior, which is new as regards even the recent past, means that through the buying and consuming of books, clothes, records, and entertainment, culturally diversified and often conflictual cultural models enter into the family circle.

Probably the young are not yet given sums of money they can use freely. A few young people married and living together with their parents are still provided what is needed for small expenses and extravagances.

The young people questioned do not seem to suffer an excess of authoritarianism or repression within the relation they have with their parents; only a few (8–14 percent) think they are kept under excessive control.

A large part of those interviewed, however, say they are not satisfied with their relations with their parents, especially the father.

On average, one in four says their relations with their parents are superficial-difficult-bad; another quarter, that they are only fair; so there are at least 30 percent living in families, who get money and personal freedom, but who judge they have poor relations with their parents. Later, perhaps, there will emerge elements to explain this feeling.

A large part of those questioned say they do not frequent a peer group (over 47 percent of the males and over 58 percent of the females). This is a focal point of the study which unfortunately does not lend itself to checking on the basis of the overall data. We have just seen that this numerous population of students belonging to relatively secure families from the standpoint of economic resources (in the later adolescent phase, from fifteen to nineteen) has a link with the parents which in terms of affective and formative relations is not satisfying for many of them. A certain correspondence between the frequencies of the responses can stimulate some hypotheses (not presently verifiable against the available data). The necessity for the adolescent to seek his own role in the peer group is not confirmed. Rather, he would seem to find himself faced with a substitute situation: where there are deficiencies in the family group the adolescent turns to the peer group.

This is a very demanding hypothesis. It should be rigorously checked against the results that would follow from the strengthening of the role of the family regarding the young, who by law can work as employees (at fifteen), can marry without parental consent (eighteen), have the right to vote (eighteen), and can enroll as volunteers in the armed forces (eighteen; shortly for females too).

Those youths who do not see themselves in a group have someone to confide in: a boy or girl friend (40–44 percent) or a partner (17–18 percent). More rarely it is the parents.

This speaking confidentially, tête-à-tête, is not hindered by traffic, noise, or the presence of other people. The discussion is so intimate that one can meet in the street (40 percent), in public places, even in discos (30 percent). Less selected are those places where someone can keep a check.

The replies to some questions provide information on the cultural background of the young respondents.

Many adolescents say they have religious ideas. Only one male in five and one female in seven say they have no religion. Of the 20,000 respondents only 960 were not Catholic by religion. However, among the Catholics the nonpracticing prevailed. Despite this, the males (69 percent) and especially the females (77 percent) had decided to take religion at school and had arrived at that choice by personal conviction (69 percent).

As regards politics there was greater interest among the adolescent males (40 percent) than their young female cohorts

(23 percent). However, these are minority choices. Even with the circumscribed, precise questions—like position on nuclear energy—there were broad areas of silence. On the referendum against the keeping of nuclear power stations in Italy, over 50 percent did not reply. Among the rest the antinuclear position prevails, especially among females, but this is always a very restricted view. More limited still is interest in environmental or ecological associations.

There is little political or cultural information, and little on current affairs. Newspapers, weeklies, and books circulate but male adolescents read mostly about sports (40 percent) and entertainment (10 percent), while females read about fashion (40 percent) and entertainment (25 percent).

The attitude toward sports is also rather weak. If many males (63 percent) and females (54 percent) say they practice a sport, large numbers cannot say what sport it is.

We now come to the heart of the study which focuses on private relations, with special reference to sexual relations and experience.

Seventy percent of males and forty percent of females come to sexual maturity without having been prepared to understand the modifications taking place in their bodies. Sex remains one of the hardest problems adolescents must face in our society and they often find themselves alone at the important transitions. The small number of males who had awaited with some awareness their first ejaculation owed this to a friend's or mother's advice or the reading of an article. The 60 percent of females prepared for their first menstruation owed this to their mother's advice. The figure of the father is almost wholly absent. With whom are the first discussions of sexuality held? In the private sphere, primarily with friends and with the mother both in the case of young males (50 percent) and young females (54 percent). Books and periodicals help (13 percent). But once more the figure of the father is absent. The school's role is marginal, like the church's.

The replies to the questions referring to sexual life and that of the couple are difficult to interpret: they are distinguished by sex but not by the age of the respondents. The many no's recurring in all the questions may be due to the very young ages of many of the respondents. So, given the ambiguity of the no, one can try to interpret the positive answers.

Females more than males tend to have a stable partner. In sentimental relations, males and females seek above all affection-friendship-love (62–71 percent). Young males more frequently stress sexual need as the basis of the relation (23 percent), and they also seek protection, a sense of security (19 percent). If we interpret these two types of attitude as indicators of a certain kind of traditionalism, they may provide us with other keys: the male wishes to take the initiative to start a relation (79 percent); the female expects him to take the initiative (64 percent); females say they are faithful (80 percent) and require fidelity (87 percent). Young males claim faithfulness (76 percent) more than they feel able to provide it (59 percent). A jealous feeling is also strong for 62 to 75 percent, but it is also true that as these males say they are seeking affection, love, and understanding in the relation, it may also be understandable that they cling to it to the point of wanting to defend it against anyone who might challenge it.

It seems possible to conclude that from their first adolescent experiences there is an attempt to satisfy a group of complex needs. There is not just the response to sexual drive or curiosity. One might say the males tend to overcome the separation between love and sexuality.

The data from the analysis let us go no further. The question thus remains open, whether one is in the presence of something new as regards, for instance, Marcuse's research. Some aspects of that analysis are still valid: family and extrafamilial structures make young men and women arrive at sexual maturity unprepared. But the social structure always calls for indiscriminate sexual consumption, to have "everything at once." One should ask if it is illusory despite this to hear in the respondents' replies an effort to live the sexual relation as a complete relation of communication with another person. Is there in them that sexual poverty so many have written about and discussed, or is there rather an attempt of a different kind?

There is a distance in the current study, betwen affective, couple-type relations, not devoid of sexual stimuli, and complete sexual relations. One repeats that the age group thirteen to nineteen is the numerical summation of situations of different kinds.

Complete sexual relations are achieved around age sixteen to seventeen, both by males (47 percent) and females (55 percent). In 30 percent of cases, they take place at thirteen to fifteen, in others at eighteen and above. In all, in the study, only 50 percent of males and 34 percent of the females say they have already had complete relations. In the face of this event, the behavior of the sexes tends to differ.

The push toward a complete relation among young males comes from love or curiosity (30–35 percent), and also from the desire to feel adult and equal to others (20 percent).

In females, love prevails (59 percent), accompanied by curiosity (20 percent) or submission to the partner (16 percent). In most cases the females have their first complete relation with a partner they consider their fiancé (60 percent), while the males say they often had this relation by chance (42 percent) and less often with their fiancée (32 percent).

Sensations of shame, guilt, and fear are entwined with this first memory in 40 percent of the females although they chose their partner with more thought. This limits their pleasant memories (42 percent) which more frequently characterize the first experience of young males (67 percent). It would seem that contraceptive practices are entrusted to the males (rubber or coitus interruptus). There is little use of the pill (15 percent), and in 20 percent of the cases no precautions were taken.

The widespread unpreparedness regarding contraception is stressed by the channels of information the young turn to: friends, in the first place (30 percent), medical services (20 percent), books (15 percent), and parents last of all.

One hundred ninety-five females say they had voluntary, clandestine abortions. Sixty of them were forced into this because they were outside the time limits set by the law from the time of conception, and forty-nine others were afraid of the hospital bureaucracy and thus perhaps were afraid of exceeding the time limits. Eighty-six believed clandestinity guaranteed greater secrecy. The most significant datum, at all events, is that compared with sixty-five voluntary legal abortions there were one hundred ninety-five illegal ones.

Males and females, in view of the values of the present and concerns for the future, give the same responses and express their preferences with equal intensity. Today, the most important things for them are love (28.5 percent) and health (25.5 percent). As I noted at the start, we are taken back to the individual. All the rest follows far behind—family, success, work, economic welfare, beauty, intelligence. The greatest commitment today is applied to the construction of one's own psychophysical personality. It is to be feared that the sociocultural characteristics of the respondents have been neglected.

For the future these young people foresee they will have to worry about having a job, the necessary basis for their own emancipation as adults (47 percent). And they want peace (23 percent). Is there a search for the obvious or a tension regarding essential needs?

In conclusion, despite their still uncertain, allusive expressions, it seems one may conclude that among young Italians there is an undoubted, if unfocussed, unease. This unease, clearly expressed at the individual level in psychological terms, has nonetheless a structural foundation and seems indeed to reflect the contradictions typical of a relatively modern industrial society, intent on accelerating the processes of maturation, especially through mass information, while at the same time denying adolescents and young people the material means for fully realizing their autonomy. From this point of view, it seems certain that adolescents pose a serious challenge to the institutions of primary and secondary socialization. But precisely in the phase where these should give evidence of their highest degree of flexibility, inventiveness, and imagination, perhaps for a conditioned reflex of defense, they close in on themselves and are resistant to any demands for information, participation, and new conceptions and practices in the power to decide which arise from youth society.

In these conditions it should not be too surprising to see the secessionist attitude of adolescents and young adults. In particular, their inability to plan their own future lives rationally should not be a surprise. The lack of historical consciousness, of a sense of the past, in favor of an immediatism which ignores everything about tradition and takes refuge in total ignorance for fear of having to measure up to difficult tasks, makes its price felt in terms of indifference or hidden anguish. The perception of time appears mutilated. Time itself is denied. But we are time. I have already said so. It is what these pages mean. Time is our very life, essentially a happening in time.

Ignorance, or worse, indifference toward the past deprives modern adolescents of the sense of values. The mummified institutions which should guard them hide or devalue them. For values, or the persons and things that have worth for us, are given value by our memory. Memory is an instrument which is essentially valuing. But the things we remember are not thereby values in themselves. They are values only because they point to, embody, as it were, experiences that are significant for us. Thus there is no eternity of values *a parte rei*. Values should not be made mythical. Their eternalization, far from guaranteeing them, distances them from the very basis of their value, puts their matrix at risk, even though this always remains the real experience of mortal human beings.

In the mass society of the throwaway, where everyone uses everyone and is in turn used, memory introduces the dimension of *pietas*, reawakens a capacity for compassion. It gives weight to a criterion of judgment on the basis of which one does not use and throw away certain things. There are things and people, situations and moments we do not let fall into oblivion, and which are for us a value. For this, we remember them, keep them before us. They do not, in our eyes, belong to the perishable ephemeral. They are unique things and persons, irreducible to anything else, not replaceable, not interchangeable. They have a hard center which lasts through time. What is it? Authenticity? Autonomy? Continuity? I have already said, but it is worth repeating, *pietas* and lamentation explode when someone dies. A prototype has broken up, something unique and unrepeatable, which once was and is no more and never shall be. Augustine said tears are the blood of the soul. It is not the life time that finishes, but the life, that life, that reaches the end.

Why do positive religious moralities remain so abstract, so asceptically extraneous to the ambiguity of the everyday, its nature which, rather than problematic, is approximate, chance, plastic, and yet extraordinary? Because they have put their values under a bell-jar, to preserve them intact, save them. They have removed them from life. But they have thus made them insignificant to living men, reduced them to empty formulas, of necessity repeated but not experienced. The crisis of the five great historical universal religions basically derives from the fact that, in order to preserve their values from all contamination, they have put them in a vacuum. But in this way they have inevitably impoverished them, dessicated them, and distanced them from effective collective behavior. To save the values they have separated them from life, put them aside from the corrosive flux of time. Thus they have ritualized them, cancelled them as rules of life at the risk of transforming them into fragile alibis for mass hypocrisy.

NOTES

1. See John U. Nef, *La naissance de la civilisation industrielle et le monde contemporain* (Paris: A. Colin, 1954), pp. 18–19; see also my *The Myth of Inevitable Progress* (New York: Greenwood Press, 1986).
2. See V. Castronovo, "Il signor spaccaminuti," *La Repubblica* 29 (Aug. 1985), emphasis added.
3. W. Helpbach, *L'uomo della metropoli* (Milano: Comunita, 1960).
4. See the exemplary book by Gianni Toti, *Il tempo libero* (Rome: Editori Riuniti, 1956).
5. A. Giddens, *A Contemporary Critique of Historical Materialism* (Berkeley: University of California Press, 1981), p. 12.
6. K. Marx, *La miseria delle filosofia*, Italian translation (Rome: Edzioni Riuascita, 1958), p. 23.
7. Giddens, *Contemporary Critique*, pp. 165–66.
8. The theoretically definitive criticisms of such misunderstandings are to be found in the work of Antonio Labriola, the only "professor" the self-taught B. Croce recognized and respected as such. He was perhaps the only Italian Marxist to have enjoyed an international reputation and influence, as we see in Trotsky's *Autobiography*.
9. J. Pieper, *Leisure, the Basis of Culture* (London: Faber & Faber, 1953). On the same wavelength and basically in the same world of thought, though of a different sociocultural stamp, see T. S. Eliot, *Notes toward a Definition of Culture* (London: Faber & Faber, 1951).
10. C. Mongardini, "Il problema del tempo nella societa contem-

poranea," paper presented at the seminar on the problem of time, sponsored by the University of Cracow, 16–20 Oct. 1986.

11. Cf. my "Nostalgia dell autorita," in *In nome del padre* (Rome–Bari: Laterza, 1985).

12. See M. Heidegger, *Sein und Zeit*, especially the whole second part concerning the ontology of the *Zeitlichkeit* or temporality.

13. Ibid. For the concept of historical horizon, see my *La storia e il quotidiano*, pp. 124ff.

14. See M. Heidegger's reflections in *Was heisst Denken?* (Tübingen: Niemeyer, 1954), esp. lesson 4, p. 2: "Memory initially signifies the interior disposition of man, and his devotion.... Disposition has a wider meaning than that common in modern speech; it means... the essential being of all human nature.... In Latin it is called *animus*, as distinct from *anima*.... Memory, in the sense of human thinking which remembers, lives where everything that offers nourishment for thought is kept in safety." On modern flattening, see M. Perniola, *Transiti* (Bologna: Cappelli, 1986).

15. Cf. my *La storia e il quotidiano*, pp. 5–9. The German movie director Wim Wenders has remarked that in the United States

> people are victims of the overweening power of TV. Video does not expose reality, it manipulates it. The TV image is made up of many lines, which have no relation to the things focussed on. It is a sophisticated trick. It does not let one really penetrate the world. Aside from Manhattan, which is not America, TV dominates daily life 24 hours out of 24. Literature, painting and theater count for nothing. Newspapers themselves have an ever-decreasing importance.... There is no more information, no more possibility of expression, there are no more ideas. (interview, *L'Espresso*, 30 Mar. 1986)

Wenders' originality is certainly not earth-shaking. That television is manipulative and that, contrary to the cinema, it writes its images on the face of the viewers is well known. However, his statements confirm a problem worth examining. According to Walker Percy, author of *The Thanatos Syndrome* (New York: Farrar, Straus & Giroux, 1987), "a large part of people's anxiety and depression today comes from watching TV" (*Times Book Review*, 5 Apr. 1987, p. 22).

16. I purposely use the Machiavellian term *effective*. It adequately indicates a research purpose which comes under the area of critical realism.

17. I use, and draw attention to, my "La macchina come 'nuova arma' Riflessioni su un autore minore," *La critica Sociologica* 73 (Spring 1985).

18. M. Morasso, *La vita moderna nell'arte* (Turin, 1904), p. 243.

19. M. Morasso, *Uomini e idee del domani—l'Egoarchia* (Turin, 1898), p. xi.

20. B. Romani, *Dal simbolismo al futurismo* (Firenze, 1969), p. 113.

21. M. Morasso, *La nuova arma (la macchina)* (Turin, 1905), p. 5.

22. M. Morasso, preface to Jean du Taillis, *Da Pechino a Parigi in automobile* (Turin: Elli Bocca, 1907), p. ix.

23. Ibid., pp. ix–xii.

24. See esp. M. Morasso, *L'aspetto meccanico del mondo* (Milan, 1907).

25. I make use, in this regard, of my article "Tempe, storia e valori in T. S. Eliot," *La critica sociologica* 78 (Summer 1986).

26. See S. Serpieri, "Il significato di 'Gerontion' nella poesia eliotiana," *Studi Americani* 5 (1954): 219.

27. S. Spender, *Eliot* (New York: Fontana-Collins, 1968) pp. 63–64. In this connection, see C. Riks, *T. S. Eliot and Prejudice* (Berkeley: University of California Press, 1989).

28. S. Serpieri, "Il significato di 'Gerontion,'" p. 226.

29. See the remarks by G. Melchiori in the introduction to the Mondadori edition of *Ulysses*.

30. Spender, *Eliot*, p. 65.

31. See R. B. Kaplan, *Notes on T. S. Eliot's Major Poems* (Toronto, 1977), p. 55.

32. G. *Williamson, T. S. Eliot* (London, 1955), p. 109.

33. T. S. Eliot, *Selected Prose* (London: Faber & Faber, 1975), p. 38.

34. *Tradition and Individual Talent*, *Selected Prose*, p. 40.

35. In this respect, see Leonard Krieger, *Time's Reasons* (Chicago: University of Chicago Press, 1989), esp. pp. 11–23.

36. Spender, *Eliot*, p. 9.

37. Ibid., p. 69.

38. S. Serpieri, "Il significato di 'Gerontion,'" p. 231.

39. Eliot, *Selected Prose*, p. 44.

40. Cf. Erik H. Erikson's contrasting position.

41. The testimony of a contemporary autobiography provides an interesting confirmation; see Bobby Greene, *Be True to Your School* (New York, 1987): "like restoring a cracked and faded old photograph."

42. The age corresponding to 1945—year of liberations: my personal one (from bureaucratic and family constrictions) and Europe's (from fascism).

43. See Agnes Hankiss's remarks in *Ontologie du Moi. Le réarrangement mythologique de l'histoire de vie*, paper presented at the Uppsala ISA congress, Aug. 1978, cited in my *Storia e storie di vita*, p. 99. See also M. Kohli, ed., *Soziologie des Lebenslaufes* (Frankfurt:

Luchterhand, 1978); C. Saraceno, *Corso della vita e approccio biografico* (Trento: University of Trento, 1986). Some interesting problematic points raised by Saraceno will in due course be taken up and examined.

44. See J. Bensman and R. Lilienfeld, *Between Public and Private* (New York: Macmillan, 1979), esp. pp. 65ff; see E. A. Shils, *The Torment of Secrecy* (Chicago: University of Chicago Press, 1956) for a treatment of the problem in a general political key.

45. D. Levinson, *La struttura della vita individuale*, in C. Saraceno, ed., *Eta' e corso della vita* (Bologna: Il Mulino, 1986), p. 128, emphasis added.

46. See my *Manuale di sociologia* (Rome–Bari: Laterza, 1986), esp. pp. 116ff. But the classic work where the family is seen as a barrier against the pressure of the purely utilitarian logic of the market is M. Horkheimer, ed., *Studi sull'autorita' e la famiglia* (Turin: UTET, 1974) or *Studien über antozität und familie* (Paris: Alcan, 1936).

47. For a remarkable testimony in support of this, see my *Colloquio con Lukács* (Milan: F. Angeli, 1983). The lively sociojuridical and moral discussion now going on in the United States regarding "surrogate mothers" who accept being impregnated and becoming bearers of the seed of a married man, with regular contract and payment, is a confirmation of this.

48. This is true not only for individuals but for the overall system. Cf. the statements of the entrepreneur Carlo De Benedetti, chief manager of Olivetti in Ivrea: "Technology is destroying space and time, the last two barriers for the human being.... In the third industrial revolution, the actors are victims at the same time.... All industries today must solve an impossible economic equation. How to liquidate the greater cost of development of new products whose life is ever shorter" (interview with P. Revzin and L. Colby, *Wall Street Journal*, 7 Apr. 1987).

49. Erik H. Erikson, *The Life Cycle Completed* (New York: Norton, 1982), pp. 56–57.

50. Erik H. Erikson, *Identity: Youth and Crisis* (New York: Norton, (1968), p. 169.

51. For an initial summary of the various positions, see D. Goleman, "Perception of Time Emerges as Key Psychological Factor," *New York Times*, 30 Dec. 1986.

52. Ibid.

53. This should not be surprising. Marx wrote just after the inquiries, mostly very accurate, organized by Her Majesty's government, and further used his friend Engels' direct testimony as a Manchester industrialist.

54. Cf. the work of this title by Karl Polany.

55. Erikson, *Identity: Youth and Crisis*, pp. 142–55. In addition to Ortega y Gasset, one should look for work on the problem of the generations in Karl Mannheim, *Sociologia della conoscenza* (Bari: Dedalo, 1974 [1928]), and in S. N. Eisenstadt.

56. For the distinction between formally codified institutions and informal institutions and movements, see my *Manuale di sociologia* p. 115ff.

57. For an instructive example of the difficulty of achieving a critical level in this regard, see among others L. Bovone, introduction to *Storie di vita composita* (Milan: F. Angeli, 1984), p. 30, n. 80.

58. See 1st ed. (Naples: Liguori, 1973).

59. The following are the salient characteristics of AIED research, about which I have been interviewed and later published a full comment in *Riforma della scuola* under the title "Con disagio verso la maturita" (Uneasily Toward Maturity) (Dec. 1986–Jan. 1987).

The research was carried out by the administration of two rigorous questionnaires (one for males, one for females), to adolescents from thirteen to eighteen to nineteen years of age. There were no interviews. For this reason too, some aspects of the subjects dealt with, though important, may not have been touched upon or sufficiently developed.

The duration of the study was about five months (May–September 1986). The questionnaires were distributed mainly in large and medium cities, with very slight distribution in rural areas. Distribution was effected through the network of AIED bureaus, some public clinics, and by using individual social and health workers, private individuals, and so forth. Replies were facilitated by the fact that the questionnaires, after being filled in, could be mailed in prepaid sealed envelopes. Distribution took place using schools, sports facilities, private houses, public halls, beaches, open spaces, public places, fast food outlets, takeouts, discos, and so forth.

Over 22,000 questionnaires were distributed by AIED. About 18,000 were returned. About 17,000 were fed into the research, to which must be added another 3,000 coming out of the publication of the questionnaire in *Cioe*. In all, 20,000 questionnaires were processed. Over 1,000 were excluded for various reasons (incompleteness, obviously contradictory replies, obscenity, etc.). As mentioned, the AIED questionnaire was published in the weekly for adolescents *Cioe* (no. 34, Aug. 1986), which had widespread distribution.

Females replied in greater numbers than males, showing generally greater readiness and interest in responding.

Annotated Bibliography

Agamben, G. *Infanzia e storia. Distruzione dell'esperienza e origine della storia*. Turin: Einaudi: 1978.

The crisis of the modern concept of experience; science and experience; fantasy and adventure; the theory of childhood and language; ritual and game; the conception of time in Western culture.

Ahundov, M. D. "Lo spazio e il tempo nella struttura della teoria fisica. *Scientia* nos. 5–8 (1978).

Alici, L. "La funzione della Distentio nella dottrina agostiniana del Tempo." *Augustinianum* no. 3 (1975)

An important subject, dealt with in a general way by contemporary authors such as Paul Ricoeur.

Alquié, F. *Desiderio di eternita*. Rome: Il Pensiero Scientifico, 1977.

The affective rejection of time; the passions; memory; habit and remorse; necessity and renunication of the eternal.

Altieri, L., et al. *Tempo di vivere. Nuove identita' e paradigma giovanile dopo il 1977*. Milan: F. Angeli, 1983.

The crisis of politics and modes of aggegation in youth groups; political identity and social conflict; the new relation with work; precarious work and expressiveness; union behavior; the organization of free time; small groups, clans, and bands.

Annual Conference of the Association for the Social Study of Time. Vol. 3, Dartington Hall, mimeographed, 1986.

Professor Francesca Brezzi and Dr. Giovanna Ambrosio collaborated in the preparation of this bibliography.

Adam, "Time in Science and Society"; Adams, "Pyschic Tomorrows"; Birchall, "Work-ending, Well-being, and Time-management"; Hantrais, "Time and Family"; Hillman, "One Western European Time Zone"; Jahoda, "The Treatment of Time in Psychology"; Latham, "Dimensionality of Event in Art"; Levine, "The Pace of Life across Cultures"; Mayr, "Social Time as Experience"; Minors, Waterhouse, and Young, "The Internal Clock, the Environment and Social Factors"; Schuller, "Serving Time: 60,000 Hours at Ford's"; Seifert, "History of Working Time in Germany"; Tabboni, "Young People and Time"; Wallman, "Time, Identity and the Experience of Work."

Ariotti, P. "Celestial Reductionism of Time. On the Scholastic Conception of Time from Albert the Great and Thomas Aquinas to the End of the 16th Century." *Studi internazionali di filosofia* (1972): 4.

Assunto, R. "Città—paesaggio—industria. Tre immagini del tempo." *Cultura* no. 2 (1971).

In this article special attention is given to time as history and time as consumption.

———. "Note sul tempo nella poesia." *Archivio di Filosofia* (1973): 1.

Attali, J. *Storie del tempo*. Milan: Spirali, 1983.

The rhythm of the sacred, the calendar, the water clocks; striker of the hours and municipal tower; clockmakers and astronomers; rural time and urban time; the use of the chronometer; precision machines; quartz watches and electronic calendars.

Bachelard, G. *La dialectique de la durée*. Paris: Puf, 1950.

The author of the "regional rationalisms" applies some basic ideas from his epistemology to the category of time.

Balandier, G. "Gli spazi e i tempi della quotidianita." Interview in *Inchiesta* no. 61 (1983).

Belloni, M. C., ed. *L'aporia del tempo*. Milan: F. Angeli, 1986.

Bergson, H. *L'évolution créatrice*. Paris: Alcan, 1907.

A work of exceptional importance.

Berti, E. "Una recente indagine sul rapporto fra essere e tempo in Aristotele." *Rivista di filosofia neo-scolastica* nos. 1–2 (1971).

Bettetini, G. *Cinema e tempo*. Milan: Vita e Pensiero, 1978.

Theory of the cinema and time; the temporal organization of filmic statement; significance of film; temporal iconism and perception; the reading of audio-visuals.

———. *Tempo del senso. La logica temporale dei testi audio-visivi*. Milan: Bompiani, 1979.

Boella, L. "Il tempo elastico in Ernst Bloch." *Aut Aut* nos. 179–80 (1980).

Bohmer, H. *Der junge Luther*. Leipzig: Gotha 1925 (1940 ed.) (5th ed. 1952).

As is well known, the basic text for the so-called *Turmerlebnis* is the fragmentary autobiographical testimony now in the preface to Luther's *Opera omnia*. Luther places his discovery of the concept of the *justitia Dei* in 1519, just when he arrived in the cell of the monastery tower. In reality almost all scholars agree that this is a chronological error, and place the date of the discovery in 1512–1513. In fact, already in his first university course on the psalms, he constantly uses the term *justitia Dei* from Psalm 35 in the sense of divine mercy.

This passage, in all its laconic character (Joachimsen termed it as of "incomparable value") is of exemplary clarity. The justice of God and thus the biblical notion of God's justice, should not be interpreted in the sense Luther (wrongly) believed to be traditional, that is, as justice God *has*, distributive (active) justice. Rather, it is justice he gives, justice we sinners receive (passive). The formula *passive justice of God* replaced for Luther the primitive one he employed of passive justification of God. This formula is considered by H. Strohl as "less heavy but fuller and more expressive," and it contains a double meaning. God is justified by us, man is justified by God. For the first meaning, Luther felt the influence of Staupitz, his superior and predecessor in the chair of Bible Reading. A well-balanced man, he played an important role in the crises of the tormented monk. A follower of Egidio Romano (St. Egidius) and Thomas Aquinas, Staupitz gives much space to grace, which inspires supernatural energies and by not imputing sins, accepts on the other hand the smaller merits of man. The unconditional recognition of one's own sins, the correct knowledge of the true meaning of repentance were for Staupitz, as later for Luther, the starting point for divine election. The *desperatio sui* became the necessary step for the path of salvation. The verse from Proverbs 18:17 was dear to Staupitz: "He that pleadeth his cause first seemeth just." This was to become a basic argument for Luther, precisely in the first meaning of passive justice. For the second meaning (man is justified by God), we find indications in the comments on the psalms and the Paul's letter to the Romans, and we find ourselves right in the Lutheran concept of *iustitia Dei iustitia fidei* (cf. Scholium to the Epistle to the Romans,

Weimar 56, pp. 169ff.): the justice of God as his quality and activity, as against the justice of men, taught by human doctrines (that is, by those of the philosophers). Luther again states that the justice of God comes to man only through faith (*iustitia Dei fit ex fide totaliter*). The cause of salvation is thus divine justice. On this problematic one should recall the interpretations of Denifle, the Catholic scholar, according to whom no Catholic theologian before Luther had understood the justice of God in a punitive sense, but that all had understood it in this passage as the justice God gives through grace. Miegge, however, stresses that Luther wanted to emphasize only that the doctors had accustomed him to understand the justice of God in a philosophical sense, and thence arose his difficulties. For Miegge, furthermore, the doctors are identified with the systematic theologians whom Luther had studied up to then, foremost among them being Gabriel Biel in whom the term *justice* is used in the sense Luther interpreted it. Two further interpretations seem to me interesting, that of Vogelsang and that of Erikson. (Vogelsang believes Luther's insight attributed to the revelation of the tower is dramatically present in the first lessons on the psalms (1513), as has been said. However, Vogelsang goes deeper, taking from Luther's text statements which are literal quotations from earlier scholars, and further brings out the swell of Luther's statements, by insisting on an archaeological study of the text itself, that is, the stratifications of his thought as they appear from the preparatory notes for the lectures, the transcriptions, and the later additions. Vogelsang also studies the type of paper and ink used by Luther, notes changes in handwriting, and analyzes the variations in the importance Luther himself attributed to different parts of his notes, variations marked by underlinings and notes of self-congratulation. Vogelsang ends by stating that in these pages by Luther we find traces of a spiritual cyclone which he experienced in having to interpret the passage *in iustitia tua libera me*. Moreover, Luther himself, now advanced in years, was to recall how, faced with this passage, he underwent an inner conflict (a spasm of horror), in which Vogelsang again makes us note the violence and passion of the language.

Erikson, on the other hand, carries out a historical, psychoanalytic study of Luther as a young man (his relations with his father, his neuroses, and his creativity). Backdating the experience of the tower to 1513 allows one to understand how at age 30, an important age for exceptional people who have had a prolonged crisis of identity, Luther's theology emerges in its entirety from the fragments of his totalistic revaluations. The comments on the psalms which concerned Luther include David's accusations against his enemies and hence against the enemies of God (Ps. 30). Luther had enemies at Erfurt, but there was also a more hidden enemy, the father who, crossed in his own plans for his son, had put a curse on the latter's spiritual life. "In the context of Martin's struggle for justification, a struggle which included the emancipation of his obsessive conscience from the jealous father and the liberation of his thought from medieval theology, this new insight on the justice of God could not, psychologically speaking, be experienced as a real solution without his hatred having found an outlet" (p. 205). Thus, concerning the event of the tower, Erikson states this is associated in Luther's mind with an earlier period of deep depression during which he foresaw an early death.

Bonessio Di Terzet. "Uomo e tempo in J. Joyce." *Giornale di metafisica* nos. 5–6 (1975).

Bonicatti, M. "Chronos-Tyche-Melancholia." *Archivio di filosofia* no. 1 (1973).

Bosserman, P. "Some Interpretations on the Dynamics of Time in Industrial Society." *Society and Leisure* no. 1 (1975).

Bottiroli, G. "Il tempo e la creazione dell'opera d'arte." *Rivista di Estetica* nos. 14–15 (1983).

Bovone, L., ed. *Storie di vita composita*. Milan: F. Angeli, 1984.

Braudel, F. "Storia e scienze sociali—la 'lunga durata'." In *Scritti sulla storia*. Milan: Mondadori, 1973 (Paris, 1958).

———. *Civilta' materiale, economia e capitalismo*. 3 vols. (Turin: Einaudi, 1981. (Paris 1979).

———. *I tempi della storia—economia, societa', civilta'*. Bari: Dedalo, 1986.

This work is arranged in three parts; the first contains a series of testimonies and interviews of an autobiographic kind; the second is a series of essays on the ancien regime; and finally there is a section on the dialogue between the different social sciences, where the historian is concerned with some of the great historians and sociologists of yesterday and today, with geography and sociology, and interscientific projects for analysis.

Brecher, P., and T. Costello. *Tanto peggio, tanto peggio. La lotta quotidiana in tempi difficili*. Turin: Rosenberg & Seller, 1979.

Brose, H. G. "From Life Time to Time in Life. New Ways of Experiencing Time in Biographies." In *Biography and Society*, presented at the Eleventh World Congress of Sociology, New Delhi, 20 August 1986.

Bruton, E. *Storia degli orologi*. Novara: Ist. Geogr. De Agostini, 1980 (London, 1979).

Butera, F., and G. Della Rocca. "Qualita' della vita di lavoro e tempo di lavoro; il caso del part-time." *Sociologia del Lavoro* nos. 17–18 (1982–1983).

Cacciari, M., ed. *Crucialita' del tempo. Saggio sulla concezione nietzschiana del tempo*. Naples: Liguori, 1980.

Canobbio, F., and A. Codelli. "Husserl: la fenomenologia della coscienza temporale." *Rivista di filosofia neo-scolastica*. no. 1 (1972).

Capozzi, G. *L'individuo, il tempo e la storia*. Naples: Jovene, 1979.

Time as alienation of consciousness; the individual as existential link of consciousness and otherness; idealism and existentialism; temporalization and historicity; the primacy of time in Kant; the developments of temporalism in Heidegger and Husserl.

Carlstein, I., D. Parkes, and N. Thrift. *Timing Space and Spacing Time*. London: Arnold, 1978.

Carsetti, A. "Tempo e realizzazione del simbolo in Proust." *Archivio di filosofia* no. 1 (1973).

Castelli, E. "Il simbolismo del tempo." *Archivio di filosofia* no. 1 (1973).

Cavallera, H. A. "Il concetto di spazio e tempo in Giovanni Gentile e il senso dell'individuo." *Giornale critico della filosofia italiana* no. 3 (1975).

Cavalli, A. "Storia, vita e quotidianita' nell'esperienza, giovanile." *Inchiesta* no. 54 (1981).

This study, based on the content analysis of some interviews with young people, permits the construction of a typology of different forms of structuring time.

———. *I giovani e il tempo*. Bologna: Il Mulino, 1985.

Cesareo, V. *La societa' flessibile*. Milan: Angeli, 1985.

Indicators of flexibility; rigid and flexible time; a possible scenario of postindustrial society; three constituent elements of flexible society; rigidity and flexibility in sociological reflection; the contribution of evolutionist sociology to the study of flexibility; social complexity and individual flexibility; flexibility and work; flexibility and education; flexibility and the tertiary.

Chaix-Ruy, J. "Sacro e profano: eternita' e tempo." *Incontri culturali in Escatologia e Secolarizzazione* no. 4 (1972).
Cipolla, C. M. *Clocks and Culture, 1300–1700*. London: Collins, 1967.
Colonello, D. "Il problema del tempo nel pensiero di Husserl." *Filosofia Oggi* no. 1 (1984).
Comolli, G. "Tempo ossessivo e tempo del quotidiano." *Aut Aut* nos. 175–76 (1980).
An article on the definition of time in relation to two types of experiences. Obsessive time is first of all subjective and linked to a reality which is not represented. Time is individual and opposed to social time.
Coser, L. A., and R. L. Coser. "Time Perspective and Social Structure." In A. W. Gouldner and H. P. Gouldner. *Modern Sociology*. New York: Harcourt, Brace Jovanovich, 1963, pp. 638–50.
Crespi, F. "La paura del quotidiano." *Inchiesta* no. 61 (1983).
Crespi, P. "Narrazione e ascolto. Aspetti e problemi dell'approccio orale in sociologia." *La critica sociologica* no. 70 (1984).
Oral testimony can help to find the structure of historical time, but it is placed within a given cultural framework. Moreover, the participation of the investigator creates links of dependency within the subjects. Further, oral testimony provides a great richness of usable elements.
Crippa, R. "L'uomo tra immortalita' e futuro." *Giornale di metafisica* no. 2 (1984).
Crisci, A. V. "Filosofia della liberazione e morte del tempo. A proposito di un libro di E. Severino." *Sapienza. Rivista internazionale di filosofia e teologia* no. 3 (1979).
Cristiani, M. "Lo spazio e il tempo nell'opera dell'Erigena." *Studi Medievali* no. 1 (1973).
D'Abbiero, M. *Hegel e il tempo storico della societa' borghese*. Bari: De Donato, 1970.
Dalla Chiara Scabia, M. L. "Istanti e individui nelle logiche temporali." *Rivista di filosofia* no. 2 (1973).
D'Angelo, A. "Considerazioni sul problema del divenire in Kierkegaard." *La Cultura* no. 4 (1978).
Davies, P. *Spazio e tempo nell'Universo moderno*. Rome–Bari: Laterza, 1980.
The model of space-time structure; the Newtonian model; the general theory of relativity and the special theory; four-dimensional space-time; the asymmetry of past and future; quantum theory; the temporal evolution of the universe; the influence of space-time concepts on societies.

De Grazia, S. *On Time, Work and Leisure.* Princeton: Princeton University Press, 1965.

Desideri, F. *Walter Benjamin. Il tempo e le forme.* Rome: Editori Riunite, 1983.

The transformation of the concept of time since Hegel; neo-Kantianism; romanticism and the critique of forms; language, politics, and theological utopia; esoterism and historical materialism; intellectuals and superstructure; Benjamin and Brecht; Adorno and the relevance of philosophy; the theses on the concept of history; politics and forms.

Dorfles, G. *L'intervallo perduto.* Turin: Einaudi, 1980.

Dumazedier, J. *Sociologia del tempo libero.* Milan: Angeli, 1978.

The increase of free time; "black" work and family activities; free time in the Soviet Union and France; the demand for cultural goods and services.

Durand, G. *Science de l'homme et tradition.* Paris: Berg, 1979.

A book filled with historical and cultural references, in the true sense "polyphonic." It is a reasoned caveat against the simplifying, but also simplistic schemas of classical rationalist decalism in favor of a rich, multiple *hermetica ratio.*

Echange et projets. *La rivoluzione del tempo scelto.* Milan: Angeli, 1986.

Eizner, N. "Memoria collettiva e immaginario in un villaggio della foresta di Orleans." *Inchiesta* no. 13 (1983).

Summary of a study carried out in a village near Orléans on the subject of people's relation with their own past and the structure of collective memory.

Eliade, M. *Mythe de l'éternel retour.* Paris: Gallimard, 1949.

The transcendence of mechanical time in the expansion of an eternal, hierophanic present.

Elias, N. *Über die Zeit* (Italian translation). Bologna: Il Mulino, 1986.

The experience of what we today call time has been modified in the past, and continues to be modified in the present, in a structured and guided way. The task of this study is to define some aspects of the structure and direction of this change and indicate how an explanation can be attempted. The Cartesian-Kantian hypothesis that men possessed the idea of time as a priori synthesis is untenable, since a single life is too short for the necessary process of learning to acquire knowledge of the specific relations represented in the concepts of cause and time, and others at the same high level of synthesis. In current use, time is a concept at a high level of generalization and synthesis, such that is presupposes a very wide background of social knowledge regarding the methods of measuring temporal sequences and their regularity. In more primitive stages men did not have this knowledge available, since it requires time to develop. A difficulty one runs into in a study on time is the lack of theory of the development of abstractions, or more correctly, of the formation of syntheses. With Einstein it was discovered that time is a form of relation and not an objective flux (Newton). However, he too fed the myth of reified time and concerned himself with the problem only within physics. A critical examination of the concept of time demands on the contrary an understanding of the relation between physical time and social time, that is, the determination of time in relation to nature and society. In the framework of physics and philosophy, time appears as a concept at a very high level of synthesis. In the practice of human societies, on the contrary, time is a regulatory mechanism of coercive forces. Time is not a physical object, but the symbol of a relation established by a human group. The hours are physical continua of change. To determine time means to establish if a certain change took place before, during, or after another change. Sociologically, time has the function of coordination and integration. In ancient times it was the priests who were the specialists in the procedures of determining time. The need for an orderly and unitary calculation of time varied with the growth and decline of state units, with the dimension and level of integration of their peoples, the degree of differentiation of their chains of commercial and industrial interpenetration. The determination of time rests on the capacity of man to connect two or more sequences of continuous change. The temporal concepts of the past/present/future type cannot be applied to the physical levels of what we call nature. Outside the reference to men who experience it, the subdivision of the natural continua of change into past, present, and future is meaningless. Temporal concepts could be distinguished into "referring to structure" and "referring to experience." Concepts like now or past/present/future are characteristic of the fifth dimension of the universe—that of subjective experience, of consciousness.

The nature-society dichotomy is insufficient. It forces investigators to treat what we might call social time and physical time (time in nature and time in society) as if they existed independently of one another, and this is not possible. Nature and society are interdependent. If we analyze time, we analyze men within nature and not separate from it. The use of clocks to determine physical sequences of events began before Galileo, but he first introduced the practice. Physical time is a late derivative of social time. With Galileo the new concept of physical time corresponding to the change in the concept of nature is proclaimed. The growth of the physical sciences involves a growing dualism of the concept of (physical and social) time. Clocks are only mechanical movements of a specific kind used by men for their own ends. The determination of time has been developed over the centuries in connection with the growth of specific social demands, amongst which is the need for men to synchronize between themselves, and with nonhuman sequences of natural phenomena, the sequences of their own activities. In the philosophical perspective there have been, on the one hand, the representatives of the idea according to which time is given to men as part of the natural order (Newton). On the other are those for whom time corresponds to a universal structure of human consciousness. Those two opposite views are within the same scheme of thought which sees the concept of time as universal and shared by all independently of their knowledge and modes of life. With very few exceptions, such as Hegel and Comte, the philosophers have always stayed prisoners of the attempt to reduce observable processes to something absolutely invariable. In many sciences there have increasingly come to the fore processual models of long sequences of change in the succession of time, which direct cognitive efforts toward the progressive discovery of change in the same temporal succession. The subjective motive whereby the discovery of something eternal within all changes has such a high value for men because it lies in their fear of death. Men are figures placed in time and in space, and can be localized in four dimensions. As the fifth coordinate one must add the determination of their movement through the symbolic universe in which they live. A representative of this fifth dimension is language. Symbolic contents are always developing. Among the properties of the concept of time there is not only that of being symbol of an extended synthesis, but also that of being a symbol of relationship. The determination of time is an activity through which men put in relation among them the successive aspects of at least two sequences of movement. The current conception of time, once acquired, is so imperative it seems a natural equipment. The structural traits of the personality are often lived and summarized conceptually as natural qualities, as innate attributes. Because of this, men educated in societies regulated by time, when they meet men endowed with a different temporal consciousness, react with incredulity.

The division between individual and social psychology prevents the grasping of the inseparability of the common, social structures of the personality from the individual and particular ones. Self-regulation on the basis of time, which is found almost everywhere in the most developed societies is not a biological datum, nor a metaphysical one, but a social datum, an aspect of the social structure in development in the personality. We have slipped into an omnipresent sensitivity to time which has become part of our personality and as such taken for granted by us. For each of us the experience of time can become comprehensible only with the aid of a reconstruction of the past. In the primitive form of the experience of time, the difference between past, present, and future, and between animate and inanimate things is not so clear as ours. For men of later societies it is easy to perceive the world through concepts of a high level of synthesis because they are born in a later phase of a long sequence of societies, in the course of which social knowledge increased without anyone planning it. This study goes beyond the current level of discussion of this problem of time, and is an attempt to replace both static-systematic thought and historical reflection with a developmental sociological idea equally distant from philosophical absolutism and historical relativism. An evolutionary process is one in which human groups, often unconsciously, draw near to the solution of a problem or draw away from it. The processes of social evolution find a partial conclusion when the problem from which they have originated is resolved. The majority of historians have so far neglected to take into consideration the long-term social processes, since systematic reflection on the problems human groups have found themselves faced with, in the past as in the present, has been lacking. The perspective of the long duration, illuminating the order of the sequences in the succession of the stages of social development, permits the demonstration of how the course of development should not be imposed from outside on inert historical material, without structure. For this structure is already to be found in documentary material itself, and can be extracted from it in the form of a symbolic, verifiable representation of the process of development.

Fabian, J. *Time and the Other.* New York: Columbia University Press, 1983.

This book proposes clarifying how the anthropologist constructs his own object of analysis. Anthropologists are the here and now whereas the objects of their discourse, the primitive, savages, the underdeveloped are the there and then. The thesis is that time, like language and money, is a bearer of meanings as it can also be an instrument of power when it reflects the logic of the mode of capitalist production.

Federici, G. *Una realta' in crescendo: il part-time*, in *Sociologia del lavoro*, nos. 17–18 (1982–1984).

Filadoro, C. *Il part-time.* Milan: Pirola, 1978.

Fimiani, M. P. *Futuro logico e tempo storico.* Naples: Guida, 1974.

Fraisse, P. *Psychologie du temps.* Paris: Puf, 1967.

Fraser, J. T., ed. *The Voices of Time: A Cooperative Study of Man's Views of Time as Expressed by the Sciences and by the Humanities.* New York: George Braziller, 1966.

Gadamer, H. G. *Wahrheit und Methode.* Tübingen: Mohr, 1965.

A basic text for understanding the typical explanatory structure of the human sciences.

Gasparini, G. "Il tempo. Un percorso sociologicamente orientato." *Studi di Sociologia,* no. 1 (1986).

Gasparri, G. "Sul rapporto fra diritto e tempo." *Rivista internazionale di filosofia del diritto* no. 4 (1974).

The temporal dimension of the juridical experience.

Geymonat, L. "Spazio e tempo nella fisica contemporanea." *Bollettino del Centro internazionale di Storia dello Spazio e del Tempo* no. 2 (1984).

Ghisalberti, A. "La concezione del tempo in San Bonaventura. *Miscellanea francescana* nos. 1–4 (1975).

Giovanelli, G., and G. Mucciarelli. *Lo studio psicologico del tempo.* Bologna: Cappelli, 1978.

The determination of the concept of time; the fields of the psychology of time; biological time; primitive notions of time; temporal attitudes in four black subcultures; personality and psychology of time; the concept of time in genetic psychology.

Granel, G. *Le sens du temps et la perception chez E. Husserl.* Paris: Gallimard, 1968.

Grossin, W. *Le temps de la vie quotidienne.* Paris: Mouton, 1974.

Gurvitch, G. *La vocation actuelle de la sociologie.* Paris: Puf, 1963.

Halbwachs, M. *La mémoire collective*. Paris: Puf, 1950.

Basic texts as reference points critical of Bergson's position.

———. *Les cadres sociaux de la mémoire*. Paris–The Hague: Mouton, 1975 (1925).

Hall, E. T. *The Dance of Life—The Other Dimension of Time*. Garden City: Doubleday, 1984.

The author distinguishes between Nordic and Mediterranean peoples on the basis of their different perception of time.

Heinemann, K., and P. Ludes. "Zeitbewusststein und Kontrolle der Zeit. In K. Hammerich and M. Klein, eds., *Kölner Zeitschrift für Soziologie und Socialpsychologie* no. 20 (1978).

Heinrichs-Offe-Wiesenthal. "Tempo di lavoro e conflitto socio-politico e industriale. *Sociologia del lavoro* nos. 17–18 (1982–1983).

The study of relations between a reduction in working time or other new forms of organization of labor, and the potential socioeconomic conflicts in the face of the crisis shaking capitalist societies.

Heller, A. *Sociologia della vita quotidiana*. Rome, 1975.

Husserl, E. *Per la fenomenologia della coscienza interna del tempo* (*1893–1917*). Milan: Angeli, 1981.

Brentano's theory of the origin of time; analysis of consciousness of time; the degrees of constitution of time and temporal objects; phenomenology of the internal consciousness of time; objective time; temporal objective and phenomenology of objectivation.

Ichino, P. *Il tempo della prestazione nel rapporto di lavoro*. Milan: Giuffré, 1984.

Isambert, F. A. "Henri Humbert et la sociologie du temps." *Revue française de sociologie* vol. 20 (1979).

Izzo, A. "Darwin e la sociologia oggi." *Quaderni dell'Istituto G. della Volpe* (1983): 7–39.

Jacobelli Isoldi, A. M. "Durée et image. Il problema del tempo in Bergson." *Il Cannocchiale* nos. 1–3 (1981).

Jacques, E. *The Form of Time*. New York: Crane Russak, 1982.

Javeau, C. *Les 24 heures du belge*. Brussels: Institut de Sociologie, 1970.

———. "Sur le concept de vie quotidienne et sa sociologie." *Cahiers internationaux de sociologie* vol. 68 (1980).

The well-known Belgian sociologist defines the basic ideas of his research, intended to free him from a crudely paleopositivistic attitude, but not thereby cloudily irrational.

———. *Le petit murmure et le bruit du monde*. Brussels: Les Epéronniers, 1987.

The remarkable gift of this short book lies in the demonstration of the untenability of scientistic positions in sociology. It is not restricted to polemical statements. It tries worthily, with an essentially positive outcome, to combine rigor and a critical sense of the problem, hinging on the idea that the individual though variously conditioned by the environment is still able through the most diverse strategies to react and, as it were, cut himself out a specific life-destiny.

Jedlowski, P. *Il tempo dell'esperienza. Studi sul concetto di vita quotidiana.* Milan: Angeli, 1986.

Johnson, T. *Natura e misura; il fiume nel tempo.* Milan: Feltrinelli, 1979.

Science and the measurement of time; methods and units of measurement; natural phenomena and biological rhythms; scientific apparatus and instruments; Einstein and the theory of relativity.

Kaplan, M. "Il tempo libero e il processo generale di teoria pratice La Polonia: studio di un caso." *Lo Spettacolo* vol. 29, no. 1 (1980).

The author, starting from the case of Poland, examines the relation between free time, ideology, and political activity.

Kiray, M. B. "The Concept of Time in Rural Societies." In *Time and Science.* Paris: Unesco, 1979.

Kolaja, J. *Social System and Time and Space. An Introduction to the Theory of Recurrent Behaviour.* Pittsburgh: Pittsburgh University Press, 1969.

Koselleck, R. *Die vergangene Zukunft.* Frankfurt am M.: Suhrkamp, 1979.

The critique of the enlightenment concept of progress which anticipated the future to the point of consuming it.

Koselleck, R., and Stempel Wolf-Dieter, eds. *Geschichte, Ereignis und Erzählung.* Munich: Wilhelm Finke, 1973.

Koyré, A. *Del mondo del pressapoco all'Universo della precisione*, Turin: Einaudi, 1967.

The book comprises two articles published in *Critique* in 1948. The first, *I filosofi e la macchina*, discusses the reasons and possible explanations for the absence of machinery in the ancient world, especially in Greece. Koyré summarizes the sociopsychological interpretation of this phenomenon argued by Pierre-Maxime Schuhl, in an article reprinted as an appendix to the book, for whom the technical stagnation of antiquity was caused both by the very structure of ancient society and its economy (aristocratic society and economy based on slavery), and by the intellectual underestimation of the socioeconomic implications of technology. The Greeks thought of and built technical tools and instruments but did not believe them an object of theoretical elaboration, or did not believe in technology. The modern mentality is very different, and from the Renaissance to the industrial revolution developed from artisans' inventions toward operational, practical thought (though always an empirical and prescientific knowledge). But above all it developed from the progress of science (astronomy, optics, physics) new technical instruments which led to social applications. In the second article, which gives the book its title, Koyré asks why antiquity produced no Galileo, and why ancient physics did not fully develop. The Greeks were good astronomers in observing and measuring the heavens patiently and exactly, but they were bad physicists, incapable of measuring the movements of objects on the earth's surface. They did not have and did not think to build calibrated measuring instruments, machines capable of scientific experiments. The development of modern science is linked to two particular technical advances: the construction of the telescope and the first successes in measuring time (building a chronometer). With the telescope a connection was established between astral physics and terrestrial physics; with the chronometer the basis was established for the development of dynamics (the science of moving bodies).

"To measure time it is indispensable to use a phenomenon which embodies it in a suitable manner; that may correspond either to a process which takes place in a uniform manner (constant velocity), or to a phenomenon which while not uniform in itself, periodically reproduces itself in its identity (isochronic repetition)." Galileo and later Huygens were oriented toward the second solution (the oscillation of the pendulum). But this discovery did not occur empirically, by observing the balancing of the great chandelier at Pisa, but thanks to the mathematical study of the fall of heavy bodies along the chords of a vertically positioned circle.

"The history of chronometry offers us an impressive example, perhaps the most impressive of all, of the birth of technological thought which progressively penetrates and transforms thought and technical reality itself." The process outlined by Koyré is: (1) the precise study of moving bodies; (2) the construction of instruments of measurement; (3) the proclamation of technological thought; (4) construction of exact machines; and (5) industrial applications (paleotechniques in the age of steam and iron).

———. *Dal mondo chiuso all'universo infinito.* Milan: Feltrinelli, 1984.

Nicolas de Cues and Marcellus Palingenius; Copernican astronomy and the new metaphysics; Kepler and the rejection of infinity; Galileo and the discovery of fixed stars; Descartes and the new mathematical cosmology; absolute time and space in Newton; Raphson and the divinization of space; Newton and Berkeley; Leibniz and Newton.

Krieger, L. *Time's Reasons.* Chicago: University of Chicago Press, 1989.

Kubler, G. *The Shape of Time. Remarks on the History of Things.* New Haven: Yale University Press, 1962.

Lalli, P. "Immaginario, simbolico e vita quotidiana." *Inchiesta* no. 61 (1983).

Landes, D. S. *Storia del tempo. L'orologio e la nascita del mondo moderno.* Milan: Mondadori, 1984 (1983).

The book is divided into three parts. "Finding Time" faces the question of how and why an invention as fruitful as that of the clock should be born in Europe and have remained a monopoly there for almost five hundred years. It transformed the Continent from a vulnerable outpost of Mediterranean civilization into a hegemonic, aggressive power. "Chronometry became at once the signal for a rediscovered creativity and a catalyzing principle of the use of knowledge as a function of riches and power." The second part, "Marking Time," examines the history of the development of the clock and of how, in the span of five centuries, one should have passed from the medieval town clocks to precision watches, from the public measurement of time to the private. The third part, "Making Time," is the longest, and talks of those who have "in effect built timepieces and clocks and the way they have done so." This concerns a treatment of economic history, of the history of the changes taking place in the techniques of manufacture of methods of production.

Lefebvre, H. *Critica della vita quotidiana.* Bari: Dedalo, 1977.

Le Goff, J. *Tempo della chiesa e tempo del mercante.* Turin: Einaudi, 1977 (Paris: 1956).

———. *Storia e memoria.* Turin: 1982.

These are articles published by the author between 1977 and 1982 for the *Enciclopedia Einaudi*, preceded by an original preface on the concept of history today.

Le Roy Ladurie, E. *Tempo di festa, tempo di carestia. Storia del clima dell'anno mille.* Turin: Einaudi, 1982.

The historical study of climate; the methodologies and techniques of dating; the chronology of climate; recent seasonal fluctuations and the phenomena connected with them; the problems of the little ice age; the movement of the Alpine glaciers and temperatures; the historical data of the medieval climate; the relations between climatic history and human history; agricultural meteorology; crops and forests; variations in harvest dates.

Lugarini, L. "Tempo e concetto. Sul rapporto tra storia e logica in Hegel." *Rivista di filosofia* nos. 2–3 (1981).

Luhmann, N. "Die Knappheit der Zeit und die Vordringlichkeit des Refristeten." In *Politische Planung*, 2d ed. Westdeutscher Vg., 1975, pp. 143–65.

Lynch, K. *What Time Is This Place?* Cambridge: MIT, 1972.

Macioti, M. I., ed. *Oralita' e vissuto.* Naples: Liguori, 1986.

Maffesoli, M. *La conquista del presente. Per una sociologia della vita quotidiana.* Rome: Ianua, 1983.

———. "Vita quotidiana e socialita'. Il superamento dell' individuo." *Inchiesta* no. 61 (1983).

———. *La connaissance ordinaire.* Paris: Les Méridiens, 1985.

A book of great lucidity and firmness in clarifying the disadvantages of the imperialism of formal logic for the social sciences. The author, professor at the Sorbonne, has developed a precise critique regarding traditional methods of social analysis.

Man and Time. Papers from the Eranos Yearbooks 3, Bollingen Series 30, New York, 1957.

Marcenaro, P., and V. Foa. *Riprendere tempo. Un dialogo con postile.* Turin: Einaudi, 1982.

The experience of the worker's labor; the ideological and political differences between generations among workers; the models of private life; the relation between working time and free time; political system and social movements.

Marramao, G. *Potere e secolarizzazione. Le categorie del tempo.* Rome, 1983.

Martins, H. "Time and Theory in Sociology." In J. Rex, ed., *Approaches to Sociology. An Introduction to Major Trends in British Sociology.* London: RKP, 1974.

Masciandro, F. *La problematica del tempo nella Commedia*. Ravenna: Longo, 1976.

Cosmic time and historico-salving time; evil as negation of the present; valorizing time in purgatory; time and eternity in paradise.

Mazzarella, P. "Creazione, partecipazione e tempo secondo San Tommaso d'Aquino. *Studia Patavina* no. 2. (1982).

Melges, F. T. *Time and the Inner Future*. New York: Wiley, 1982.

Mercure, D. "L'étude des temporalités sociales. Quelques orientations." *Cahiers internationaux de sociologie* vol. 67 (1979).

Merton, R. K. "Socially Expected Durations: A Case Study of Concept Formation in Sociology." In W. W. Powell and R. Robbins, eds., *Conflict and Consensus: A Festschrift for L. A. Coser*. New York: Free Press, 1984, pp. 262–83.

Midion, A., and J. L. Jackson, eds. *Time, Mind and Behavior*. Berlin: Springer, 1985.

Mongardini, C. *Il problema del tempo nella societa' contemporanea*, presented at the University of Cracow seminar, Oct. 1986.

Moore, W. E. *Man, Time and Society*. New York: Wiley, 1963.

Morris, R. *Time's Arrows. Scientific Attitudes toward Time*. New York: Simon & Schuster, 1984.

Munari, A. "Tempo e spazio nella psicologia genetica." *Bollettino del Centro internazionale di Storia del Spazio e del Tempo* no. 1 (1985).

Nelson, G. K. "Time in Development Studies: A Convergence of the Dialectic and Phenomenological Thought." *Genetic Psychology Monographs* no. 2 (1983).

The current study of time changes within the developmental disciplines has shifted its attention from the specific qualities of time to the characteristics of sequential order, with the consequence of viewing time as independent of the subject, reducing it to nothing more than absolute time, an antiquated model, unsuited to the relational study of behavioral processes. This cognitive separation of the contents of experience (space) from time distorts the space-time framework, or phenomenological base, from which all individuals and groups of people have to work out their development and history.

Nicolini, T. "Teoria e prassi nella conoscenza del tempo." *Incontriculturali* nos. 3–4 (1977).

Nole, L. *Tempo e sacralita' del mito. Saggio su C. Lévi-Strauss*. Rome: Bulzoni, 1981.

Diachrony-synchrony and opposition between history and

structure; Lévi-Strauss and history; rejection of humanism and traditional historicism; the unconscious and the human spirit; the sacred between the divine and the human; primitive thought and the structure of myths.

Novotny, H. "Time, Structuring and Time Measurement: On the Interrelation between Timekeepers and Social Time." In L. Fraser, ed., *The Study of Time*. New York: Springer, 1975.

Ong, W. *The Presence of the Word*. New York: Simon & Schuster, 1970.

Ossola, C. "Sospensione del tempo." *Archivio di filosofia* no. 1 (1973).

Paci, E. "L'istante." *Aut Aut* no. 136–137 (1973).

A note on the philosophical idea of the instant.

Pancaldi, G. "Spazio e tempo nella teoria darwiniana." *Rivista di filosofia* no. 1 (1973).

Paolicchi, P. *Esperienza del tempo e realta' sociale*. Pisa, 1976.

Time in psychology; genesis and individual differentiation of the temporal horizon; social psychology of time; development of the personality and social structure.

Paolucci, G., ed. *Il disagio del tempo*. Rome: Ianua, 1986.

Parkes, D., and N. Thrift. *Time, Spaces and Places*. New York: Wiley, 1980.

Parodi, M. *Tempo e spazio nel Medioevo*. Turin: Loescher, 1981.

Neoplatonism; time and movement in Plotinus, Augustine, and Scotus Ehrigen; time and space; Thomas Aquinas; logic and mathematic in the Cxiv; the thought of William of Ockham and John Buridan; developments in mathematics and physics.

Pellegrini, M. *Tempo, pianificazione e struttura. Sistemi urbani Rivista quadrimestrale Scienza, Citta' e Territorio* no. 3 (1983).

Analysis of the concept of social planning in Luhmann.

Piaget, J. *Lo sviluppo della nozione di tempo nel bambino*. Florence: La Nuova Italia, 1979.

Time and movement in the child's view; the order of events and the duration of intervals; succession of perceived events and simultaneity; perception of duration and the transitiveness of time relations; measurement of time and the idea of age.

Pomian, K. "Introduzione ad una storia filosofica del tempo." *Bollettino del Centro internazionale di Storia dello Spazio e del Tempo* no. 1 (1983).

The different dimensions of time; psychological, natural, liturgical, political, and physical.

———. *L'ordre du temps* Paris: Gallimard, 1984.

Pomticelli, L. "La coscienza della presenza del prima e del poi." *Rivista di filosofia neo-scolastica* no. 2 (1975).

Prigogine, I., and I. Stengers. *La nuova alleanza. Metamorfosi della scienza.* Turin: Einaudi, 1981 (Paris, 1980).

Pronovost, G. "La structure et la signification des temps de loisir dans le temps de la vie quotidienne." *Loisir et Société* no. 2 (1982).

Reeves, H. *L'evoluzione cosmica.* Milan: Feltrinelli, 1982.

Reichenbach, H. *Da Copernico a Einstein. Il mutamento della nostra immagine nel mondo.* Rome–Bari: Laterza, 1985.

The Copernican image of the world; the astronomical problems of movement and gravitation; the theory of electricity and the observations on the movement of light; the special theory of relativity; the relativity of movement; space and time.

Ricci, U., et al. *Vivere e progettare il tempo La prospettiva temporale nel comportamento umano.* Milan: Angeli, 1985.

The conception of time in psychological and philosophical studies; the concept of duration and succession; temporal experience and cognitive and motivational factors; the concept of temporal perspective; temporal perspective, identity, and sociocultural context; adolescence and work; temporal perspective and sickness.

Ricoeur, P. *Temps et récit.* Paris: Seuily, vol. 1, 1983.

The author analyzes the relation between the narration of a story and the temporal character of human experience. Time becomes human experience. Time becomes human time to the extent that it is articulated into a story. On the other hand, the tale achieves its full meaning when it becomes a condition of temporal experience. Time is an aspect of the movements of the universe. If there were no one to count the intervals there would be no time. The activity of the tale lies in constructing temporal wholes—configuring time.

Roberts, D. D. *Benedetto Croce and the Uses of Historicismo.* Berkeley: University of California Press, 1987.

Robin, C. *L'imaginaire du Temps Retrouvé. Hermétisme et écriture chez Proust.* Paris: Minard, 1975.

Sabbadini, A., ed. *Il tempo in psicoanalisi.* Milan: Feltrinelli, 1979.

The unconscious and time; the concept of time as defense of the integrity of the ego; time in psychoanalytic technique; the psychopathology of time; the psychology of temporal perception; the subjective perception of two aspects of time—duration and atemporality; time as dimension of affects.

Samuel, N. "Loisir, valeur et structure symbolique des temps sociaux." *Loisir et Société* no. 2 (1982).

Schlick, M. *Spazio e tempo nella fisica contemporanea. Una introdu-*

zione alla teoria della relativita' e della gravitazione. Naples: Bibliopolis, 1983.

The Newtonian concepts of space and time; the principle of special relativity; the geometrical relativity of space; the mathematical formulation of the relativity of space; movement, inertia, and gravitation; the postulate of general relativity; the finiteness of the universe; space and time in philosophy.

Sebastiani, M. "Husserl lettore de 'Sein und Zeit': la questione del tempo e il problema della fenomenologia." *Il Cannocchiale*. nos. 1–3 (1981).

The problem of phenomenology and Husserl's notes on being and time; transcendental phenomenology and phenomenological ontology; the "present" and atemporal consciousness; the problem of temporalization.

Sedlmayr, A. "La morte del tempo." *Archivio di filosofia* no. 1 (1973).

Sesti, M. "La duplicazione del tempo. Appunti su Kant e il primo Heidegger." *Il Cannocchiale* nos. 1–3 (1981).

Sexl, R., and H. R. Schmidt. *Spazio Tempo. La relativita' ristretta*. Turin: Boringhieri, 1980.

Sheenan, T. J. "Tempo ed Essere nelle lezioni di Heidegger del 1927." *Filosofia italiana* no. 3 (1978).

Sirianni, V., and M. Eayrs. "Tempo e lavoro: razionalizzazione, flessibilita' e ugughlianza." *Rassegna italiana di sociologia* no. 4 (1985).

Sorabji, R. "I concetti di spazio e di tempo nell'antichita' classica e nel pensiero cristiano antico." *Bollettino del Centro internazionale di Storia dello Spazio e del Tempo* no. 2. (1984).

Sorokin, P. A. *Sociocultural Causality, Space, Time*. Durham: Duke University Press, 1934.

Sorokin, P. A., and R. K. Merton, "Social Time: A Methodological and Functional Analysis." *American Journal of Sociology* vol. 42, no. 5 (Mar. 1937).

Soule, G. *Time for Living*. New York: Viking, 1955.

Spaventa De Novellis, L. "Il tempo libero nell'utopia di Tommaso More." *Revue internationale de sociologie* no. 2. (1971).

———. "Ancora del tempo libero: note agli interrogativi e alle riflessioni di una studiosa." *Lo Spettacolo* no. 3 (1978).

Stagl, J. "Vom Dialog zum Fragebogen." *Kölner Zeitschrift für Soziologie und Sozialpsychologie* no. 31 (1979): 611–38.

Stame, N. "Sociologia in profondita', lunga durata e movimenti sociali." *Inchiesta* nos. 63–64 (1984).

An analysis of the critical observations by Braudel on Gurvitch's sociological theories. Part of a special number devoted to Braudel.

Stella, G. "L'interpretazione temporale del diritto di G. Husserl." *Rivista internazionale di Filosofia del diritto* no. 4 (1980).

Steri, B. "Pascal e le diverse strutture della temporalita' nel 'divertissement', nella 'machine' e nella 'grace'. *Giornale di metafisica* no. 3 (1981).

Tabboni, S. *La rappresentazione sociale del tempo*. Milan: Angeli, 1984.

Social time and natural time; time in preindustrial societies; time and industrial society; "time budgets"; time and complex society; time as a structuring social element and as total cultural fact.

———. *Tradizione e coscienza storica. Il tempo di ieri e il tempo di domani*. Genoa: Ecig, 1984.

Tabboni, S., ed. *Tempo e societa*. Milan: Angeli, 1985.

Time and social interaction; the multiplicity of social times; times and models of rationality; the binding nature of the deadline; consciousness and control of time; temporal perspective and social structure; social expectations of duration.

"Temps et société." *Revue internationale des sciences sociales* no. 107 (1986).

Time in the historicosociological perspective; society and the concept of time; social change and the modification of the concept of time; history and the sociology of free time in France; the dimension of social time as an index of the distance between classes in Italy.

Totaro, F. *La costituzione della temporalita. IKON*, nos. 82–83 (July–Dec. 1972).

Turco Liveri, G. "Tempo e liberta' nell'eterno ritorno di Nietzsche." *Il Cannocchiale* nos. 1–3 (1981).

Valenti, A. *Il problema intertemporale nell'ordinamento comunitario europeo*. Milan: Giuffré, 1977.

Succession of juridical norms through time; fait accompli and rights requested; temporal succession in community law; state laws.

Ventura, A. "Il tempo nel pensiero di G. Bachelard." *Rivista di filosofia neo-scolastica* no. 1 (1984).

Vitiello, V. "Nietzsche e la crisi del 'tempo-immagine'." *Il Cannocchiale* nos. 1–3 (1981).

Whitehead, A. N. *Il concetto della natura*. Turin: Einaudi, 1975.

Whitrow, G. *The Natural Philosophy of Time*. New York: Harper & Row, 1963.

Yaker, H., H. Osmond, and F. Cheek, eds. *The Future of Time. Man's Temporal Environment.* New York: Doubleday, 1972.

Yates, O. *The Art of Memory.* Chicago: University of Chicago Press, 1966.

Zadra, D. *Il tempo simbolico. La liturgia della vita.* Brescia: Morcelliana, 1985.

Zambelloni, F. "Tempo e eternita' nell'estetica di Schiller." *Filosofia* no. 3 (1973).

Zellkind, I., and J. Sprug, eds. *Time Research: 1172 Studies.* Metuchen, N. J.: Scarecrow, 1974.

Zenther, H. "The Social Time-Space Relationship: A Theoretical Formulation." *Sociological Inquiry* no. 1 (1966).

Zerubavel, E. *Ritmi nascosti. Orari e calendari nella vita sociale.* Bologna: Il Mulino, 1985 (University of Chicago Press, 1981).

Already the author of a study on the use and perception of time in a hospital, Zerubavel here offers a theoretical framework for what he defines as the sociology of time. This is an area of sociology still being defined, but which has fruitful heuristic potential which emerges where one makes an insertion, in a temporal perspective, of the theory of models and cultural values, the theory of knowledge and religion, ideology, the identity of social groups, and so on. Time conditions and regulates the life of the individual. It spans the moments of cohesion, places barriers and limits to subjective action, and confines behavior to the standards and perspectives of the group. Pregnancy and times of sexual intercourse, holidays and meals, the timetable of lectures, and so on, imply sequences and deadlines which live in the order of temporal succession. The first and most important institution that man has invented to establish and maintain temporal regularity is undoubtedly the calendar. This is basically responsible for the majority of temporal models whereby all societies, institutions, and social groups try to regulate the order and sequence of their own actions. Then, on the calendar there is superimposed the timetable or the introduction of microspans which regulate the behavior of individuals and groups within society. Two thousand years ago, the rabbis regulated individual and collective life with a rigid daily calendar of religious services. But it is with the coming of the Benedictine monasteries that there arrived the rigid scansion of the hours (*horarium*) which, since the Middle Ages, is characteristic of the rationality of modern everyday life. The author examines more specifically the concept of calendar, showing how time forms are connected with the formation of groups and their consolidation (in-group feeling), and that of sacred and profane time. The final chapter on private time and public time starts from Simmel's proposals on the multiple participation of the individual in groups and networks of relationship which represent the sociological connections relevant for defining the individuality-group relation. According to Zarubavel multiple participation, which he explains as segmented participation, involves the individual in a frantic network of social relations which force him to take on ever-different roles and functions in which, however, he is only partially involved. The institution of privacy as private time sanctions an individual time that is essentially "socio-fugal," in the sense that—like certain socially defined spaces—it is deliberately designed to prevent or at least discourage the formation of human contact and to separate individuals. Public time, on the other hand, is essentially sociopetal, in the sense that it deliberately wishes to promote the institution of human contact and gather people together."

Index

About the Author

FRANCO FERRAROTTI is Professor of Sociology and Chairman of the Ph.D. Program in Social Science at the University of Rome, Italy. He is the author of *The Myth of Inevitable Progress*, *Five Scenarios for the Year 2000*, and *The End of Conversation* (Greenwood Press, 1985, 1986, 1988). His previous English-language works include *An Alternative Sociology*, *Max Weber and the Destiny of Reason*, and *A Theology of Non-Believers*. His articles have appeared in numerous publications, including *Social Research*, *State, Culture, and Society*, and *La Critica Sociologica*.

www.ingramcontent.com/pod-product-compliance
Ingram Content Group UK Ltd.
Pitfield, Milton Keynes, MK11 3LW, UK
UKHW020207290926
15470UKWH00001B/51

9 780313 268281